Don’t Burn It Here

Don't Burn It Here

Grassroots Challenges to Trash Incinerators

Edward J. Walsh

Rex Warland

D. Clayton Smith

The Pennsylvania State University Press
University Park, Pennsylvania

Library of Congress Cataloging-in-Publication Data

Walsh, Edward J. (Edward Joseph)
Don't burn it here : grassroots challenges to trash incinerators / Edward J. Walsh, Rex Warland, D. Clayton Smith.

p. cm.
Includes bibliographical references (p.) and index.
ISBN 0-271-01663-9 (cloth : alk. paper)
ISBN 0-271-01664-7 (pbk. : alk. paper)
1. Environmentalism—United States—Case studies.
2. Incinerators—Government policy—United States—Citizen participation—Case studies. I. Warland, Rex H.
II. Smith, D. Clayton. III. Title.
GE197.W35 1997
363.72′85—dc21 97-10419
CIP

Printed in the United States of America
Published by The Pennsylvania State University Press,
University Park, PA 16802-1003

It is the policy of The Pennsylvania State University Press to use acid-free paper for the first printing of all clothbound books. Publications on uncoated stock satisfy the minimum requirements of American National Standard for Information Sciences—Permanence of Paper for Printed Library Materials, ANSI Z39.48-1992.

We dedicate this book to all working for genuine democracy,
and especially to

Rody, Joani, and Marty
Carol, Brett, and Mari
Hammer, June, and Lena Mae

Contents

Preface

A casual conversation during an August 1988 vacation about a new type of trash plant becoming increasingly popular across the nation sparked the research reported in the following pages. Between sips of beer on a Sea Isle, New Jersey, beach under the late afternoon sun, vacationing brothers exchanged small talk about the past year as they waited for their children to finish showering before supper. One of the Chicago-based environmental consultant's passing comments about his experiences in the Midwest was surprising: "It seems that the more democratic the processes at these waste-to-energy siting projects from the start, the more likely they are to be successful." While this had not been true for the seemingly comparable nuclear power enterprise with which his older brother was more familiar—"secrecy" and "tyranny" rather than "democracy" were words more commonly associated with that industry—it was a fascinating notion deserving closer study.

A seed grant from The Pennsylvania State University's Center for Research in Conflict and Negotiation provided resources to prepare a proposal for external support. A subsequent grant from the Ford Foundation's Fund for Research on Dispute Resolution would not have materialized without essential feedback from the above-mentioned consultant, Jim Walsh, who was as quick to offer useful criticisms of proposal and chapter drafts as he was to elbow under the boards his more gentlemanly older brother in one-on-one basketball.

The numerous local conflicts emerging around this new method of handling municipal waste provided an opportunity to test and repair contemporary social movement theory. Our previous work on grassroots mobilization in the wake of the Three Mile Island accident had suggested continuities as well as disjunctions between social movements precipitated by suddenly imposed grievances, such as nuclear accidents, and the older and more widely studied protest processes emerging among such deprived collectivities as African Americans or women. These protests over the sitings of modern incinerators promised to provide multiple comparable instances of increasingly common but less-studied protests against new technologies.

We came across a number of grassroots groups from at-risk communities successfully challenging political and economic elites, although we did not find a positive relationship between democratic processes and successful sitings. Ordinary people increasingly confront large bureaucracies over which they feel they have little or no control, but the evidence in the following chapters reveals that corporate power is more fragile than many citizens imagine. Despite multimillion-dollar budgets and claims to scientific truth, some new technologies are shortsighted and have serious flaws that are checked only by the mobilization of countervailing grassroots political power. In response to demeaning corporate accusations that they are NIMBYs (Not-in-my-backyarders) or BANANAs (Build-absolutely-nothing-anywhere-near-anythings), increasing numbers of citizen challengers define themselves as legitimate defenders of a common good being threatened by corporate shortsightedness and greed.

We also found enough evidence to caution against any headiness about U.S. society's having achieved a balance between corporate and grassroots power. In the course of our research, we encountered too many people who were intimidated by societal elites and modern technologies. Perceiving themselves, families, or friends as vulnerable to economic retaliation, they became immobilized by fear and/or cynicism. Some were wearied and/or confused with the exaggeration and name-calling by proponents as well as opponents in these sometimes heated disputes. Corporate sponsors of such technologies too often arrogantly ridicule the alleged ignorance of grassroots opponents, while some of the latter caricature the former as greedy, diabolical incarnations rather than as businesspersons attempting to find financially profitable solutions to challenging technical and political problems. In the cases examined here, each side insisted that the other had prostituted "science" by using intentionally bogus data in support of its own economic and political agenda. After hearing such charges and countercharges, some people in the surrounding areas said—more often with actions than in these specific words—"A plague on both your houses" and went about living their own private lives without becoming involved.

Yet the evidence against the wisdom of such public passivity continues to accumulate. Serious recycling of municipal wastes was not even a significant option in the United States until a few years ago. As a result of the large number of grassroots struggles, the recycling alternative to incineration became increasingly widespread, so that whereas less than 10 percent of municipal solid waste was recycled nationally in 1970, more than 24 percent was being recycled by 1994. Things had turned around so much, in fact, that by the early 1990s counties in the northeastern United States were concerned about running out of trash for their recently built incinerators and were hiring computer operators, trash-enforcement officers, and even judges to run the equivalent of a traffic court for rubbish violators, to ensure that local "flow control" laws were being followed and that refuse was not being hauled out of county to cheaper disposal sites (Stranahan, 1995). County officials worried that such practices would deprive local governments of the materials necessary to keep their waste-to-energy (WTE) incinerators operating at or near peak capacity and cause the county to default on its bonds or to raise taxes to pay off its debt. In 1994, the U.S. Supreme Court stunned these officials and the waste business in general by ruling such "flow control" practices unconstitutional and by labeling trash a free-market commodity like any other. This landmark ruling disturbed not only county waste officials who had large investments in modern incinerators, but also Wall Street investors who worried that the bonds issued for such projects were suddenly at risk. Predictably, there was litigation challenging the applicability of this Supreme Court decision in specific cases. Thus, within the space of two decades, the municipal waste worries of many county authorities had changed from "We'll need a new multimillion-dollar incinerator to handle our growing mounds of trash" to "We'll have to monitor our county's valuable trash more closely because it is essential for feeding our multimillion-dollar incinerator." The grassroots struggles analyzed in the following pages, and similar struggles across the United States, played major roles in this dramatic transformation.

Social scientists often implicitly assume that political processes are not only influenced but virtually determined by existing arrangements and demographic patterns. They portray individuals and groups as mere pawns in larger systems, with little power to transform the political landscape. In the wake of both successful and unsuccessful insurgencies, social analysts hasten to explain why what happened was the inevitable outcome of existing conditions. Delightfully, our fieldwork data in Chapters 4 through 7 offer little support for such structural reductionism, revealing as they do the impact of individual ingenuity and grassroots creativity in the creation of social change. Would the loosely coordinated national anti-incineration movement ever have coalesced if the chemistry professor from northern New York State who appears in a number of our cases had not become involved? Would the proposed incinerator in Cape May, New Jersey, have been defeated if a substitute schoolteacher had given up in the face of seemingly insurmountable odds? Similar questions about idiosyncratic characteristics may be asked about most of our case studies—so much so that some informed observers cautioned us against seeking any typical models across sitings and defeats. We persisted, however, and the patterns we eventually identified were more subtle than the literature suggests.

Those studying human behavior are even less able than quantum physicists to avoid influencing the processes they observe and report upon. An ironic example of an incinerator industry study that came back to haunt its sponsors with unintended consequences figures prominently in the following chapters, and we also acknowledge the unwanted attention our own survey attracted in one ongoing conflict. Although most were extremely cooperative, some activists refused to participate, and a few even attempted to subvert our efforts. We have tried to minimize the influence of our own political partialities in the following pages, revising some generalizations in the light of comments on earlier drafts from informed individuals on both sides of the various disputes. Readers who do not want to wait may turn immediately to the two concluding memos to county officials and grassroots activists at the end of this book in their efforts to unmask our remaining conscious (and unconscious?) biases.

Activists and elites, no less than researchers, too often demonize opponents because of a variety of fears or in the interest of self-aggrandizement. We have tried to neutralize such tendencies by using multiple cases, multiple methods, and a wide variety of informants. Some industry decision-makers and activists who, for different reasons, prefer to repress outside analyses of their activities—the former in order to control interpretations with their own spin, and the latter to protect themselves from any advantage such knowledge may bring political and economic elites—may object to this book. We hope, however, that it contributes to more genuine democratic decision-making in a wide variety of comparable community projects.

Methodologically, we rely upon both quantitative and qualitative data in our efforts to explain why some of these projects were sited in the face of organized citizen protests while others were defeated. Our survey items were considerably improved by the thoughtful suggestions of two colleagues, Bob Herrmann and Art Sterngold, who also helped iron out some glitches in the administration of the survey by a marketing firm in northern

New Jersey. Graduate students Kevin Ross and Glynis Daniels made important early contributions—Kevin to the Philadelphia fieldwork, and Glynis in getting our Lackawanna County newspaper file data into accessible form. Eileen McCaffery did outstanding work for us in assembling the Philadelphia newspaper file. Audrey Glover was very helpful in keeping us abreast of the ongoing Broome County (New York) dispute and its surprising conclusion. Scores of other individuals assisted us in understanding and analyzing events in the various counties, and we have attempted to acknowledge each of these with footnotes in subsequent chapters. If we have failed to mention someone, we apologize for the unintentional oversight. Scores of graduate and undergraduate students in Ed Walsh's social movements course also offered useful criticisms as well as helpful repairs to earlier drafts. Robbie Swanger's cheerfulness in preparing and revising our tables made that work seem like fun, and we also thank Jan Fleischmann for her labors on the index. Peter Potter and Peggy Hoover, both with Penn State Press, were terrific to work with as they prodded us to make this book more reader friendly. We especially thank Jim Jaspers for his many valuable comments on a previous version of our manuscript.

The three states focused upon in this book—New York, New Jersey, and Pennsylvania—are major players in the municipal solid waste story for the United States. New York and New Jersey are among this nation's largest waste exporters, and Pennsylvania is its major waste *importer*. We are not aware of any significant influences that the different political structures in these states may have had upon the processes here considered, but their similarities in terms of both industry and anti-incineration modes of operating were remarkable. Indeed, some of the same industry representatives and anti-incineration activists appeared at different sites across state lines to argue for and against proposed projects. The overwhelming majority of proponents and opponents we encountered were not notably different from ourselves, our colleagues, or our students. Readers will decide whether the analysis of their political struggles in the following pages is fair. We hope it also contributes to the enhancement of social science, and especially to more genuine grassroots democracy.

Introduction

When first proposed in the United States during the 1970s, waste-to-energy (WTE) incinerators appeared to be ideal solutions to the growing mounds of trash in our "throwaway" society. Promising to convert useless garbage into electricity while saving precious landfill space, trash incinerators seemed perfectly timed to respond to a national need. Within a decade, however, a grassroots anti-incineration movement emerged as a vibrant offshoot of the environmental movement. Why and how did this change occur? Do social science theories about grassroots protest mobilization explain this dramatic evolution in perceptions of incinerators from panacea to poison? Alternatively, do the specific details of these resistance processes contribute anything to social science theories?

Throughout the 1970s and into the early 1980s, as citizens produced increasing volumes of trash and the available space for solid waste landfills throughout the United States decreased, county officials often decided behind closed doors to build what are variously called trash-to-steam or waste-to-energy plants.[1] Such facilities incinerate trash and use some of the energy created in the combustion process to generate electricity. During this period, plans for these million-dollar projects were seldom jeopardized by the sporadic local opponents objecting to such trash disposal burdens as the additional truck traffic, another smokestack in the neighborhood, obnoxious smells, and threats to property values. Special compensation benefits, including tax relief, which were part of the normal contract usually created

1. A secretary typing one of our early taped interviews had never heard the phrase "trash-to-steam plant," so she had to guess at what the speaker using the phrase quite frequently was saying. One Monday morning, we received an otherwise flawless written transcription that initially puzzled us with its multiple references to a "trashed esteem plant," prompting Martin Walsh to draw a wilted potted flower as our project's logo.

more than enough project supporters during these early years to neutralize the few outspoken opponents.

Proponents of these facilities insisted there was no connection between such compensation packages and health risks. They wanted to eliminate any public perception that financial incentives were presented as bribes or tradeoffs for health risks. On the contrary, they emphasized that everything possible would be done to make the WTE facilities safe, and that the whole project was a very low risk operation. Compensation, they insisted, was offered only for the additional neighborhood burdens associated with the plant. Health risks were a separate issue, and they were addressed by the county or its outside consultants in risk assessments.

After the mid-1980s, however, such issues became much more problematic as siting attempts were increasingly challenged by organized opponents. This change was due to the accessibility of critical technical information about incineration, increased enthusiasm for recycling, and to the emergence of loosely coordinated grassroots protests. The ideas and the networks comprising this anti-incineration tributary of the broader environmental justice movement only converged gradually from different sources. All of the siting projects we examine in this book were in process during the latter part of the 1980s, and each encountered some organized resistance.

Opposition forces focused upon three general areas: the fairness of the original siting decisions, health and safety issues associated with the technology, and the recycling alternative to burning. The fairness question derived from the early environmental justice movement claim that such burdensome projects were typically targeted for poorer neighborhoods, where residents had least political power. The industry's health-risk assessments were criticized by technically literate opponents as biased political documents using statistics to appear as scientific studies. And trash incineration itself was portrayed as a socially irresponsible alternative to serious recycling. Our title encapsulates these different foci of the anti-incineration movement: DON'T BURN (recycle instead) IT (unhealthy and unsafe trash) HERE (a location selected for political rather than scientific reasons).

Incinerator proponents, in response to such challenges, used both carrots and sticks. They not only ridiculed opponents as selfish NIMBYs with no concern for public trash disposal problems, but also insisted verbally on their wholehearted support for recycling programs. They defended the sophistication of both the environmental impact statements and health-risk assessments associated with incinerator sitings, and some industry consultants encouraged county officials to establish citizen advisory committees with real decision-making powers on everything from the initial site selection to other critical issues such as health, compensation, complementary recycling, and continuing monitoring of the facility. Whatever other purposes such committees might serve, they promised to be politically useful in neutralizing grassroots opposition.

Although industry spokespersons insisted that even the most successful waste recycling programs could absorb only 25 percent of the total waste stream, recycling enthusiasts put the upper limit closer to 85 percent. There are few issues with more relevance to incineration and over which the two sides are further apart. Even though informed observers in each camp now agree that the priorities for handling this nation's trash problems should be waste reduction at the source, followed by the recycling of whatever cannot be reduced, then incineration of nonrecyclables that are burnable, and finally landfilling whatever remains, serious disagreements surface when the discussion becomes more specific. It is virtually impossible for local communities, counties, or even states to work realistically on the first of these priorities—waste reduction—because it is a problem best dealt with at the national level through legislation covering, for example, bottling and packaging requirements. This means that recycling estimates are obviously critical in determining how much trash will be available in any county for incineration.

Our findings will presumably be applicable to a wide variety of state and county siting processes, because wherever problematic and costly engineering projects such as incinerators, landfills, radioactive waste facilities, and similar ventures are initiated comparable dynamics are likely to emerge. If the new generation of smaller nuclear power plants envisioned by advocates in that industry is ever actively promoted in the United States or Western Europe, for example, similar grassroots protests are predictable, and we also expect that the dynamics for siting such institutions as new prisons will resemble the processes analyzed in the following chapters.

Contemporary social movement theory predicts that characteristics among grassroots insurgents, such as dense associational networks and a strategic framing of grievances, will be critical indicators of successful protest mobilization. Analysts also commonly assume that the lower the socioeconomic status of a discontented collectivity, the less likely it will be to mount a serious collective challenge. Each siting effort considered in this book encountered some organized resistance from citizens in the host community, and our findings raise questions about such general assumptions while also illustrating the need for new models of actual protests. Confrontations between grassroots citizens and proponents of technological projects have become an increasingly common variant of environmental dispute. As of early 1997, for example, Pennsylvania had already spent more than $10 million on studies of public involvement in conjunction with what had been to that point a futile search for an acceptable site on which to build a federally mandated low-level radioactive waste facility. Careful analyses of comparable projects—such as those included in the following chapters—make not only theoretical but also economic sense.

We initially intended to focus upon four sited and four defeated WTE projects, including two of the eight where final decisions had not yet been made but which seemed likely to wind up as a successful siting and a defeated one. Things did not turn out precisely as planned, as we explain later. After consultations with individuals on both sides of the WTE controversy, we decided upon specific sites where conflicts had recently occurred or were still in process. Interviews with key proponents and opponents at each site helped us understand specific idiosyncratic issues and then construct questionnaires focusing upon relevant environmental and social movement variables. A marketing survey firm administered a telephone survey to randomly selected residents living within a few miles of each project, and to whom we refer as "backyarders." The reader will find a more thorough discussion of our operational definition of backyarder and questionnaire items in Chapter 3 and in the Appendix. In addition to continuing interviews with proponents and opponents at each site while reconstructing the histories of these struggles, focus group discussions with activists belonging to the challenging groups were also conducted. These activists, as well as members of citizen advisory committees (where they existed), were asked to complete questionnaires similar in most respects to the telephone surveys. We also assembled eight comprehensive files of local newspaper articles, enabling us to document the evolution of each siting controversy.[2]

In an attempt to neutralize any biases associated with the narratives, we

2. Although there was considerable ongoing interaction across task lines among the co-authors throughout this project, the interviewing and focus group work was done by Ed Walsh, and Rex Warland concentrated his efforts on the construction, administration, and interpretation of the various surveys. Doug Smith, who joined the project as a graduate student after the data were collected, worked on both quantitative and qualitative dimensions of this research while also reanalyzing some of the data for his own excellent doctoral dissertation in sociology, "Power and Process in the Siting of Municipal Solid Waste Incinerators" (Penn State University, 1996).

revised Chapters 4 through 7 after sending early drafts to opponents and proponents at each location for their criticism and feedback. We were especially encouraged by one response, which included an unsolicited concluding comment: "These chapters should be required reading for industry consultants as well as grassroots activists before they become involved in any more such projects."

A historical sketch of both the evolution of the modern incinerator industry and the development of the corresponding anti-incineration movement is presented in Chapter 1. Then, in Chapter 2, we discuss theoretical similarities and differences between the movements of the 1960s and the technology movements focused upon here. Chapter 3 presents information from our telephone interviews with people living in the immediate vicinities of the eight incinerator sites. These survey results from backyard residents not only introduce the reader to each project but also raise important new questions, which are then addressed in the subsequent four chapters, where we draw on our fieldwork data to reconstruct each struggle in detail. The map in Figure 1 (see Chapter 3) identifies the various counties and specific sites, numbering them according to the order in which they are discussed in Chapters 4 through 7. Chapter 8 integrates the survey and fieldwork data to answer our focal questions, summarizes the theoretical relevance of our findings for the environmental justice movement, and then concludes with two hypothetical memos synthesizing central insights from our study: the first to proponents of modern incinerators and comparable technological projects, and the concluding one to citizen activists.

1

The Incinerator Siting Controversy in the United States

In 1990, the United States had 140 incinerators burning trash and generating electricity, but almost twice that number had also been canceled in the previous eight years because grassroots opposition made it increasingly difficult to build a trash plant anywhere in the nation after 1985 (Curlee et al., 1994). Before focusing on eight community struggles involving such facilities during that critical period, we trace the evolution of the technology and the increasing opposition it encountered. Even though the primary method of waste disposal in the United States is and always has been landfills, incineration has played an important auxiliary role throughout the twentieth century. Earlier versions of these facilities, however, were seldom challenged. How did things evolve thus, and why are increasing numbers of citizens aligning themselves against this modern variant of a technology that has been around in one form or another for more than a century?

First-Wave and Second-Wave Incinerators

Before the 1890s, waste disposal was handled by individuals, and burning was one of the more common methods. Municipal incineration of garbage is an extension of this trend, which has followed a cycle of boom and bust since the first garbage incinerator was built on Governors Island, New York, in 1885 (League, 1993). It is convenient to divide the history of incineration in the United States into three major waves, with the third—including our eight projects—as the one of primary interest for us in this book.

The first wave began in the United States in the 1880s and continued until the early years of the twentieth century. Precipitated by the increasing waste generated by urban populations, the notion of solid waste management developed during this period, when the options for handling it included landfilling, farm usage (animal food or fertilizer), water dumping, reduction, and incineration (Blumberg and Gottleib, 1989; Melosi, 1981). Before World War I, more than 70 percent of U.S. cities used some type of source separation program for wastes before disposal, as it became increasingly common to utilize incinerator designs then in use throughout Europe to cremate solid waste. The advent of motorized waste hauling pushed more urban areas toward combined collection and disposal systems by making such solutions more economical than separation. The first incinerator was built in the United States in 1885, and within the next twenty-five years 180 others were constructed (Whitaker, 1994). Because the country had an abundance of coal, oil, and wood to burn, this first generation of incinerators did not use excess heat to create electricity, even though pilot facilities doing that were tested during that period (Blumberg and Gottleib, 1989). Most of these incinerators, however, were abandoned by 1910 because the U.S. waste stream was much wetter than the European designs could handle and therefore required considerably more energy input to burn (Whitaker, 1994). Incinerator operators who attempted to get by without using additional fuel for the wetter waste found themselves with lower temperatures, which caused incomplete burning and increased gas as well as smoke (Melosi, 1981).

A second wave of trash plants that were better able to handle the wetter U.S. waste stream began in 1906, peaked by the 1930s, and ended only in the 1960s. Several hundred of these second-wave incinerators with British design adaptations were in operation by 1915, and between 600 and 700 had been built in urban areas by the 1930s (Blumberg and Gottleib, 1989). While landfills and dumps continued to be the primary method of waste disposal during this time, because land was cheap and such facilities were easy to construct, incineration became increasingly common in the cities as per capita waste generation increased. Between 1930 and the 1960s, the waste stream itself underwent major changes as hazardous new consumer and industrial products entered the market and as television advertising encouraged both wasteful packaging and excessive consumption. For example, as the price for plastics dropped because of increased production and use, refillable bottles were replaced by plastic one-use containers (Whitaker, 1994).

Some Early Environmental Concerns

By the mid-1960s, when approximately 10 percent of the solid waste in the United States was being incinerated, the effects of a rapidly growing and changing waste stream, combined with poor waste regulation, forced the U.S. Bureau of Solid Waste Management to label 94 percent of all landfills and 75 percent of all incinerators then in use "environmentally inadequate"—based upon the criteria of air-water pollution, insect-rodent problems, and physical appearance (Blumberg and Gottlieb, 1989). The first rumblings of what would gradually evolve into an environmental movement then began in the United States, with a small but growing number of citizens emphasizing the importance of conservation, health, and natural beauty while insisting on the reduction of waste and litter. A few voluntary recycling programs emerged, and some citizens pressured the federal government to regulate the packaging of goods (Selke, 1990; Szasz, 1994).

Between 1965 and 1970, under pressure from an increasingly vocal citizenry, the federal government passed three noteworthy laws intended to improve the management of solid waste. The Solid Waste Disposal Act (1965) had two major goals: (1) to stimulate research on methods of disposal as well as reduction of the waste stream and (2) to provide technical and financial support in planning for solid waste. One result of this legislation was the sanitary landfill, which was a definite improvement over the existing type because it had a bottom liner intended to prevent leaking water from contaminating ground water and because at the end of the day it was covered with a layer of dirt to retard odor and prevent animal and insect problems. Two years later, in 1967, came the Air Quality Act, which was subsequently amended by the Resource Recovery Act of 1970. The 1967 legislation required older incinerators to add air pollution controls, commonly referred to as "scrubbers" or "precipitators." While significantly improving air quality, these pollution control techniques were expensive enough to put most of the second wave of incinerators out of operation because of the prohibitive costs involved in retrofitting existing facilities (Blumberg and Gottleib, 1989). Then came the Resource Recovery Act of 1970, indicating a federal shift of emphasis in solid waste management away from waste disposal and toward more recycling and reuse of materials (Popp et al., 1985). It made increased federal funding available to state and local governments for new resource recovery and solid waste disposal facilities, and also promoted demonstration projects for more efficient handling of solid wastes.

The Waste-to-Energy Industry Flourishes

This convergence of federal funding, new laws, and stiffer landfill requirements—in conjunction with a constantly increasing trash stream—stimulated U.S. engineering firms to initiate the third wave of incinerators in the 1970s (League, 1993). Again, engineers in the United States looked to Europe for solutions and began to obtain the patent rights to a different variant of municipal solid waste plants. In an effort to distance the new wave from its older, dirtier predecessors, the solid waste industry coined terms such as "waste-to-energy" (WTE) and "resource recovery" (Blumberg and Gottleib, 1989).[1]

Additional factors, ranging from a global oil crisis in the early 1970s to local landfill disputes, converged to promote the emergence of these new plants. The Middle East energy crisis during those years, for example, coincided with the closing of the dirtier plants to make a new type of waste facility generating electricity as a by-product of its trash incineration particularly attractive. When the nuclear reactor market contracted about the same time, some manufacturers such as Westinghouse became actively involved in this new application, which required relatively minor adaptations of existing nuclear power plant technology. The electricity generated during those days of the first oil embargo could profitably be sold to utilities, which were then paying approximately three times their pre-embargo prices. An important related piece of federal legislation during the late 1970s that promoted the construction of WTE facilities was the 1978 Public Utility Regu-

1. Modern incinerators are typically referred to by the industry and by pro-incinerator parties as "waste-to-energy" or "resource recovery" facilities. Opponents, on the other hand, frequently refer to them simply as "trash" plants. Occasionally, both sides may use the label "trash-to-steam." Such descriptive terms are used interchangeably in this book.

latory Policies Act (PURPA), ensuring that the local utility would purchase electricity generated by such plants (Curlee et al., 1994). Relatively lax regulatory controls on emissions, and various tax incentives, also contributed to the favorable climate for manufacturers of these new plants.

Within a few years, additional firms were competing for contracts to build and operate such WTE facilities, and their construction designs usually fit into one of two categories, depending upon the amount of preprocessing done to the solid waste before incineration. Because these two different types are occasionally referred to by disputants in subsequent chapters, it is useful to distinguish them briefly here. The most typical WTE design, and the one eventually settled upon in each of the cases we analyze, was commonly referred to as "mass burn incineration." Such facilities accept mixed trash and burn it at high temperatures, removing only large and noncombustible items beforehand. A second incinerator design, introduced shortly after this mass burn model, was the refuse derived fuel (RDF) system, which separates out recyclable material and then shreds the remaining waste into uniformly sized pellets. These pellets are then incinerated much like coal (League, 1993).

Although a number of early mass burn and RDF plants experienced significant start-up and operational difficulties—for example, large volumes of ash, emission problems, excessive operating costs, and difficulties in marketing electricity—they became a bright hope during the oil embargoes because they promised to provide steam and electricity while dramatically reducing trash volume (Blumberg and Gottleib, 1989). They were viewed as welcome partial solutions to the threatening problem of municipal solid waste accumulation in the midst of declining landfill space, especially in the Northeast, the Midwest, and California. There was also decreasing public acceptance of new or expanded landfills in nearby neighborhoods, with the U.S. Environmental Protection Agency (EPA) estimating total municipal solid waste (MSW) generation in the United States at more than 160 million tons in 1986, and rising at a rate of "slightly over one percent each year" (U.S. Congress, OTA, 1989:4). In a survey during the late 1980s, for example, the EPA indicated that approximately one-third of all existing landfills were expected to close by 1994 (EPA, 1988).

After two years of stopgap reauthorizations of the Resource Recovery Act, Congress passed comprehensive new legislation in 1976: the Resource Conservation and Recovery Act (RCRA). It included federal criteria for solid waste disposal facilities and a plan to phase out all open dumps within five years. Especially noteworthy is that this legislation authorized federal grants for up to 75 percent of the cost of resource recovery demonstration projects (Popp et al., 1985). The RCRA did not, however, provide for a comprehensive regulatory role in the management of municipal wastes. Because there was not yet any aroused mass base of public opinion on the issues of either hazardous or municipal waste management, the legislators followed the typical pattern in essentially being captives of the industry they were supposedly regulating and decided "to regulate disposal rather than to encourage waste reduction" (Szasz, 1994:35). Thus tax incentives, the increasing volume of waste generated, and the shutting down of old facilities promoted the construction of new incinerators. When hazardous waste unexpectedly became a major issue shortly after the 1976 RCRA legislation was passed—primarily because of the alarming number of toxic waste sites then being uncovered—the EPA shifted most of its budget and personnel from municipal solid waste to hazardous waste (League, 1993; Blumberg and Gottleib, 1989; Szasz, 1994).

The Cresting of WTE Industry Growth

Third-wave incinerator construction crested in the 1980s. As the federal government concentrated on the more politically pressing problem of hazardous waste, the Reagan White House encouraged privatization in the municipal solid waste field. While regulatory agencies such as the EPA were underfunded, incentives for private corporations to become involved expanded. The EPA's encouragement of private sector involvement coupled with favorable tax law changes and industrial development bonds made these MSW incinerator projects very attractive to private corporations (Blumberg and Gottleib, 1989). In this climate, orders for waste-to-energy plants mushroomed.

State and local legislators grew increasingly concerned in the early 1980s about "the garbage crisis" as landfills were reaching capacity or closing due to new regulations. Public anxieties about landfill safety—especially ground water contamination and toxic air emissions—narrowed choices of disposal methods during this time when widespread recycling efforts as a means of handling the waste stream were seldom seriously discussed. County authorities turned increasingly to incineration. In 1980, for example, it was estimated there were 60 of the third-wave WTE incinerators on line, under construction, or proposed across the United States and by 1985 that total had reached 200. Yet as of 1990 there were only 202 WTE facilities "operational, under construction . . . or in the advanced stages of planning" (Curlee et al., 1994:38). Although the numbers in the various categories varied somewhat because of the criteria used, the evidence was clear that the upward curve of new WTE orders had leveled after 1985.[2]

Thus the waste-to-energy industry flourished into the mid-1980s as growing numbers of county governments viewed incinerators as solutions to their trash problems. Utilizing the energy created in the combustion process to generate electricity, WTE plants increased output with size even though the total amount of energy produced is actually small in absolute numbers because, as one industry analyst noted, "fundamentally, these projects are not energy projects but waste management projects" (Hocker, 1991:14–15).

Convergence of Problems for WTE

The challenges ultimately responsible for the waste-to-energy industry's reversal of fortunes later in the decade began emerging during the early 1980s from both external and internal sources. The drop in world oil prices during this period, for example, made outside investors less enthusiastic about the technology as an alternative source of electricity. More commonly, however, external problems were linked with industry vulnerabilities.

Loosely organized opponents were becoming increasingly common by the early 1980s. Until this time, regulatory challenges and environmental reviews of modern incinerator projects had been virtually nonexistent. Although state air permits were required, there had been no organized opposition, and the level of regulation by state agency personnel was comparatively primitive. County officials confronted with a landfill crisis simply decided where they wanted to build a WTE incinerator, put out a request for proposal, and selected a vendor from among those submitting proposals. Any additional issues were then discussed, special compensation for the host community was typically agreed upon, and costs were passed on to county

2. The "Waste Not" newsletter published by incinerator opponents Paul and Ellen Connett offers an example of different numbers prompted by variations in criteria. This source listed 102 operating WTE incinerators in 1988, 125 in 1991, and 113 in 1993 ("Waste Not," no. 251, Connett and Connett, 1993). The trend here is obviously similar, however, because of the years required for these plants to move from drawing board to operational status.

residents whose trash would be delivered to the incinerator. But things were changing.

By the early 1980s, sitings were becoming more difficult as opponents raised a host of new issues. Scientists such as Barry Commoner emphasized the harmfulness of dioxins released by the burning of trash, and the incinerator industry was accused of viewing these plants as the primary solution to the nation's trash problems rather than as weak substitutes for serious source reduction and recycling. A few state and federal regulators were also beginning to echo the challengers' criticisms. While the West Europeans and the Japanese were already using more integrated solid waste management systems, encouraging companies to recycle and reduce raw materials in production, incinerator manufacturers in the United States were still ignoring recycling as part of their waste management systems (Ross, 1992). Such major U.S. designers and builders as Ogden Martin, Wheelabrator, and ABB had only adopted WTE technology from abroad without emphasizing European and Japanese multifaceted approaches that included systematic recycling as part of an overall trash management system (Hocker, 1991).

Between 1982 and 1985, siting efforts caused much more acrimonious disputes because of the increasingly organized opposition they had to confront. This was the period when the anti-incineration movement coalesced and protests proliferated. In a 1989 interview, industry consultant Carolyn Konheim elaborated on some of the factors contributing to this change of climate:

> Tax incentives for the builders of these plants contributed significantly to the rapid initial growth of this industry during the latter part of the 1970s and early 1980s. . . .
>
> In about 1980 or 1981 there was a discovery of dioxin in the emissions of an operating plant in Hempstead, Long Island. And the fear of dioxin that was propagated from that point forward gave a legitimacy to a variety of different concerns. . . . A lot of scientific attention was given to dioxin. . . .
>
> Barry Commoner was the first to make dioxin out to be the most threatening human carcinogen that ever appeared. . . . Commoner was viewed by the public as the major guru on dioxin. . . . People associated dioxin with Agent Orange, even if only subliminally, and mixed with their natural aversion to an incinerator this did not help the industry. . . . Then you have someone coming along saying, "I'm

> giving you a better answer to your trash problems: I'm saying you can recycle everything instead!"
>
> So even though the risk from dioxin was minimal according to scientific studies, the public doesn't want to accept any kind of risk. . . . And then there were a few incinerator projects which didn't work well and caused large tax increases. These were used by opponents.[3]

Both the economic and political tides were thus becoming less favorable for the incinerator industry after 1982. We turn next to a consideration of the evolution of the anti-incineration movement in the United States that both fostered and exploited these shifting tides.

The Anti-Incineration Movement

It was essentially the combination of outside opposition with internal industry vulnerabilities which promoted the effective mobilization of grassroots challenges to the third wave of incinerators. It is convenient for our purposes to discuss the complex combination of factors contributing to the evolution of the national anti-incineration movement under four general headings: (1) problems with the technology itself, (2) an internal industry document which was leaked and used by opponents, (3) the emergence of a major anti-incineration organization, and (4) involvement by a chemistry professor who became a national opposition leader. While each had its own special importance contributing to the mobilization of a national anti-incineration movement, their mutual interactions are most instructive. Internal technical problems and the embarrassing document, for example, became serious liabilities for the industry only when organized opponents used them to mobilize grassroots protests. Readers will find examples of such blendings of factors common in the following chapters.

3. Carolyn Konheim, interview by Ed Walsh on November 26, 1989, in Brooklyn, New York. Konheim discussed the history of the industry with which she was intimately acquainted, and her insights are drawn upon throughout this chapter. She is cited only when directly quoted. Because all of the interviews and focus group discussions drawn upon in this book were conducted by Ed Walsh, we will not repeat this fact in subsequent footnotes.

Technical Issues

Backyard protesters were slow to use Barry Commoner's technical papers on the dioxin issue in the early 1980s because his technical language was virtually inaccessible to most laypeople. Not only was the content of these scientific reports difficult to grasp without the relevant background education, but they were not widely available. Early complaints from backyard citizen groups against their local incinerator projects typically focused instead upon burdens they were being asked to bear—such as increased traffic, noise from the facility, and decreased property values—for the presumed good of the larger county.

It was only when serious questions about such alleged benefits to the common good were raised, however, that effective grassroots mobilization became possible. The major environmental and health concerns responsible for the change from the routine sitings of the 1970s to the increasing challenges a few years later included atmospheric emissions, ash management, and prior recycling. These were not issues incinerator proponents wanted publicly discussed, but protest organizations also emerged during this period to popularize technical jargon and increase awareness of such negative evidence previously obscured from public view.

Atmospheric Emissions

Some familiar waste stream materials, such as papers and plastics, create furans and dioxins when trash is improperly burned. Harmful emissions also result from heavy metals contained in batteries or used in small quantities as additives in household items. Other problems derive from hazardous household wastes—estimated to comprise about 1 percent of all MSW by weight—burned without proper controls. The trace emissions from WTE plants commonly considered carcinogenic include furans, dioxins, and several metals, such as arsenic, cadmium, and chromium. Noncarcinogenic emissions include lead, mercury, particulates, sulfur dioxide, and nitrogen oxides. While all such emissions are potentially dangerous, the most serious concern has been raised about furans and dioxins, which are widely considered highly carcinogenic (Curlee et al., 1994).

Dioxin, the poison that caused the entire community of Times Beach, Missouri, to evacuate in the early 1980s, had really only begun to be studied by researchers in the early 1970s. Its presence in modern incinerator by-products became a public issue when it was discovered as one of the chlorinated hydrocarbons contained in the flue gas from the combustion of municipal solid waste (U.S. Congress, OTA, 1989:226). Commoner and others attempted to focus public attention on two large groups of compounds known as chlorinated dioxins (dibenzo-p-dioxins, or PCDDs) and chlorinated furans (dibenzofurans, or PCDFs), as well as other harmful constituents of incinerator by-products (Denison and Ruston, 1990:9). Some of these compounds are highly toxic to laboratory animals under certain conditions, and the EPA also considers one particular form—2,3,7,8-tetrachloro-p-dibenzodioxin, or TCDD—a probable human carcinogen (U.S. Congress, OTA, 1989:226). Increasing concern about dioxins in the environment and in fish and dairy products led to a 1985 moratorium on new incinerators in Sweden, and new dioxin limits were established at much lower levels (U.S. Congress, OTA, 1989:244).

Ash Disposal

Another extremely controversial environmental and health issue associated with WTE incinerators is ash disposal. Toxic ash is the noncombustible part of municipal solid waste. Critics consider it the Achilles' heel of the modern incinerator industry because the more successful a given facility is in removing toxins from the air, the more toxic its residual ash necessarily becomes. After combustion, an amorphous, glasslike material that includes minerals, metals, unburned organic carbon, dirt, and grit remain (U.S. Congress, OTA, 1989:247). Combustion results in both fly ash and bottom ash. Fly ash only amounts to between 5 and 15 percent of total ash and consists of light particles, such as that carried off the grate by turbulence. Bottom ash, on the other hand, is the relatively coarse uncombusted or partly combusted residue that accumulates on the incinerator grate. The ash from U.S. incinerators is typically 15 to 30 percent by weight and 5 to 15 percent by volume of the MSW original. The potential risks associated with this ash are the subject of considerable debate because several exposure pathways exist for pollutants in the ash. Substances may leach into ground water, and airborne or waterborne transport of ash during handling operations or from landfills may lead to inhalation, ingestion in food crops, or dermal exposure (U.S. Congress, OTA, 1989:254).

Because most of the information about released substances from modern solid waste incinerators is based upon modeling and inference rather than upon "real data on what happens to substances released from stacks," any scientific health risk assessment "should expressly acknowledge the uncertainties in the data and models" (Denison and Ruston, 1990:219). While modern air pollution control equipment can remove from emissions a significant fraction of particles and certain gases, these substances are often merely transferred to the ash, which must be transported and disposed of in some storage facility. Furthermore, it is difficult to predict the actual environmental behavior of incinerator releases because laboratory predictions often give inaccurate results:

> For instance, we now know that dioxin in the environment does not break down very quickly, although pure dioxin does break down when exposed to ultraviolet light in the laboratory. This discrepancy is explained by the fact that the dioxin released from incinerators is tightly bound to organic material and is shielded from degradation. (Denison and Ruston, 1990:220)

The primary controversy involving incinerator ash disposal is over what type of landfill should be used. About 47 percent of incinerator ash is currently sent to sanitary landfills along with other municipal waste, and the remaining 53 percent is sent to specially designed monofills. A critical question is whether incinerator ash should really be classified as a hazardous waste and thus landfilled at sites with even more stringent standards and designs, as well as at much higher disposal costs than conventional municipal landfills (Curlee et al., 1994:14).

The Recycling Alternative

"Recycling" means different things in different contexts, ranging from source separation to the construction of glass manufacturing plants to the enforced composting of household garbage. This variety of specific meanings for the term contributes to the confusion over solid waste recycling estimates, which range from a low of 15 percent by some incinerator proponents, to a high of 90 percent by enthusiasts such as Barry Commoner (1990:590). Because local communities or counties can do relatively little in the way of waste reduction involving national legislation covering bottling and packaging requirements or reductions in the use of raw materials,[4] and because the old solution, landfilling, became increasingly problematic, the choices for increasing numbers of counties and local communities boiled

4. By the early 1990s, Massachusetts had initiated an ambitious program under the Toxics Use Reduction Act, which aimed to decrease industrial-toxic-chemical waste 50 percent by 1997. Other countries were moving in this direction as well (Ross, 1992:11).

down to recycling and incineration. And although the two may theoretically be considered complementary, they typically appeared as alternatives in zero-sum contests. If anything near recycling enthusiasts' estimates of percentage reduction of the trash stream were accomplished, of course, it would leave little incentive for the incinerator industry, because the EPA estimated that the number of recycling programs in U.S. communities had almost doubled between 1986 and 1991 (Ross, 1992).

The financing arrangements for incinerators often precluded significant recycling because they required from the county a certain number of tons per day for burning. If recycling efforts subsequently reduced the amount of waste available for incineration under such contracts, counties would default on their loans or be required to import trash from elsewhere to feed their incinerators. Critics insisted that the "put or pay" clauses in such contracts that assured vendors some specified level of compensation regardless of fluctuations in the solid waste stream—and thus locked the county into a certified minimum amount of trash to be burned—served as counterincentives to the development and expansion of a serious recycling program.

The Embarrassing Cerrell Report

A 1984 industry-sponsored report from a California consulting firm suggested strategies and tactics for overcoming the increasing grassroots resistance being encountered in the early 1980s, but after opponents somehow obtained a copy this report itself became an important protest resource. Prepared for the California Waste Management Board by a Los Angeles group calling itself Cerrell Associates Inc., the ninety-page document advised potential incinerator builders on ways to overcome public opposition. While its usefulness for the intended industry audience is uncertain, this document became invaluable to incinerator opponents who commonly referred to it as the Cerrell Report. Drawing upon questionnaires received from city officials and private planners across California, telephone interviews with sales representatives of several major WTE manufacturing firms, engineers, key officials at successful as well as unsuccessful sitings, and a literature review of public attitudes toward "noxious" facilities, the report was entitled "Political Difficulties Facing Waste-to-Energy Conversion Plant Siting." Geared to assisting proponents in avoiding or overcoming public resistance, the following excerpts suggest its guiding assumptions and main themes:

> The environmental advantages of this [WTE] process are enormous: an alternative energy source in waste products reduces the amount of land that need be spoiled for landfilling purposes. . . . Most importantly, the development of an alternative energy source will help lessen the costs and dangers associated with dependence on oil and coal resources. . . . Waste-to-energy facilities are economically and environmentally progressive waste management facilities. The benefits of disposing waste products in a useful, efficient, and safe manner make traditional waste management techniques such as landfilling seem prodigal at best. . . . The most formidable obstacle to Waste-to-Energy facilities is public opposition. . . . What decisions can be made in selecting a site that encourage community acceptance of the project? . . . Candidate sites can be suggested partly on the basis of neighborhoods least likely to express opposition . . . [and] the people most likely to express opposition to a Waste-to-Energy project can be targeted in a public participation program and public relations campaign.

After discussing issues confronting proponents of various waste management projects in the face of public opposition, the authors suggest demographic features of populations where such opposition is most likely and list "for facility proponents the selection of a Waste-to-Energy site that will tend to offer the least amount of political resistance to the project" (Cerrell Associates, 1984:5). They conclude with "a program designed to enhance community acceptance of a proposed Waste-to-Energy project." The report's authors summarize their findings on the demographic characteristics of support and opposition to such facilities in an appendix and list "strong indicators" of so-called "least resistant" communities: community size under 25,000, rural location, receiving significant economic benefits from facility, politically conservative, above middle age, and where the average education is high school or less. They also include "mild indicators" of these same communities: Republicans, business or technology related occupations, low income, Catholic religion, not previously involved in voluntary associations, and longtime residents in the community.

By the second half of the 1980s this document, which was clearly not designed to flatter citizens at proposed sites, was circulating among anti-incineration groups. Activists used it, as will be seen, to mobilize local residents who resented being patronizingly regarded as "least resistant" to what opponents characterized as the incinerator industry's exploitation.

The Citizens Clearinghouse for Hazardous Wastes (CCHW)

In the early 1980s, the beginnings of what would become a major organizational player in the national anti-incineration movement emerged in the Northeast when Lois Gibbs, a Love Canal housewife, started the Citizens Clearinghouse for Hazardous Wastes. The era of routine sitings only ended as such organized opposition to third-wave incinerators coalesced in the mid-1980s to exploit the industry's technical and political vulnerabilities. Each of the projects analyzed in this book were in process during this transition period. Loosely coordinated groups and individuals were reaching out to one another for resources and moral support in challenging such projects. In the absence of such outside assistance, opponents in the early communities selected as host sites for modern trash plants found themselves unable to neutralize the industry's scientific arguments and the "NIMBY" label used to tar challengers.

Within a few years, calls from communities challenging municipal waste projects were pouring into the CCHW for assistance, and the organization responded by preparing useful booklets for local activists. Focused initially on hazardous rather than municipal wastes, this protest organization expanded its concerns in response to the demands it received for information. Andrew Szasz (1994) provides a thorough analysis of its important role in the contemporary environmental movement. Given the blurring of the boundary between certain variants of municipal and hazardous wastes associated with incinerator ash, it was a natural transition for a national citizen organization initially focused on the latter to expand into the former.

Started by the leader of the Love Canal Homeowners' Association because she wanted to "give other people the kind of help she wished she had gotten when she first started at Love Canal" (CCHW, 1986:15–16), the CCHW defined its mission as Saul Alinsky-style community organizing—"help(ing) people help themselves." One of its many documents made available to community groups, "The Polluters' Secret Plan" (Collette, 1989), used cartoons and caricatures to summarize the Cerrell Report in a way designed to arouse potential WTE host communities. It also gave abundant advice on effective countermoves that targeted communities might use. In its early pages, for example, was the claim that "not one single citizens' group that's followed our advice on how to block a proposed LULU ('Locally Undesirable Land Use,' an industry term) has ever lost" (Collette, 1989:1). It advised communities targeted by industry for a LULU to

> . . . point out that industry and government are wrong to insist "it's got to go somewhere" . . . [because] we believe this is a self-fulfilling prophecy. . . . Take a "NIABY" position—Not in Anyone's Back Yard—and take a hard-line stand FOR waste reduction, recycling and toxic use elimination. (Collette, 1989:9)

This CCHW organizing manual warns targeted communities against believing that the citizen advisory committees sometimes established in conjunction with such projects will be anything other than co-optative mechanisms: "The advisory committee process is built on the assumption a site will be built. The committee's purpose is to 'advise' industry or government on how much pollution 'the community' will tolerate" (Collette, 1989:13). The same manual warns communities against accepting "economic enticements," which it labels "bribes to get communities to accept new sites":

> Just say NO. Don't play the game by your opponents' rules and don't tolerate those who do. . . . Don't sit on the committees. Challenge and poke holes in the claims they make about how wonderful their new facility will be. Demand that they be specific about promises of jobs or new industry. Ask: "How many jobs? What skills will be required for those jobs? What will they pay? How many local people will be hired? What new companies are itching to come to this town? NAME THEM!" (Collette, 1989:14)

Running through a litany of industry and government strategies used against local community groups, the manual suggests counterstrategies that, it claims, will be more effective for grassroots groups speaking for "the true majority position," concluding:

> In the final analysis, the millions of dollars industry and government have been forced to spend in the so-far futile efforts to cope with the Grassroots Movement for Environmental Justice is a tribute to our success. (Collette, 1989:33)

By the mid-1980s, the CCHW was providing important assistance to a variety of grassroots activists. Some consider the CCHW a key player in the development of the "hazardous waste" movement into a much more broadly defined "toxics" movement that is still evolving and may become the vehicle for reinvigorating progressive politics in the United States (Szasz, 1994). The CCHW certainly played an important role in providing grassroots activists with information as well as emotional support in a number of the cases analyzed in the following chapters.

Paul Connett, Academic Activist

Around the same time that the CCHW expanded its focus to include WTE incinerators, Paul Connett, a chemistry professor at St. Lawrence University in northern New York State, was also becoming involved in the struggle against a proposed trash plant in his own backyard. By the late 1980s, if an industry proponent were asked to identify a single individual symbolizing the opposition, Connett's name would probably be the first one mentioned. Although social scientists prefer to emphasize the impact of structures on individuals, the process is often delightfully reversed in the real world. While Connett did eventually create small teams to assist in his work, many viewed him as a personal symbol of technical and scientific opposition to incineration—a hero to grassroots opposition groups and a demon in the eyes of the WTE industry.

In an interview, Connett and his wife, Ellen, discussed his involvement.[5] After receiving his college degree from Cambridge University in England in 1962, Connett worked at other jobs until he earned a Ph.D. in chemistry from Dartmouth two decades later, in 1983. His area of research was the interaction of metals with biological systems. Connett had been at St. Lawrence University in Canton, New York, for only eighteen months before becoming involved in the local incinerator controversy. He explained:

> Around Christmas of '84 the college librarian came up to me one day and said, "You're a chemist; what's the issue of incineration?" My first reaction when I heard her description of incineration was that it sounded like a pretty sensible idea—you would get rid of a lot of lousy landfills (these landfills weren't too good around here), and

5. Paul and Ellen Connett, taped interview on December 11, 1990. Direct quotations in this and subsequent chapters of our book are from this interview.

> you would concentrate the problems all in one spot which could be regulated, and you would make energy to boot. My feeling was, initially, that this librarian and the few other concerned people in this area who opposed it were overreacting to the proposed incinerator. . . .
>
> Very few local people knew much about the issue at this time, . . . although nationally Barry Commoner and NYPIRG were out there very much concerned, and some groups in California were also opposing such facilities. . . . And while the Cerrell Report was also completed about this time, I didn't become aware of it until years later. . . .
>
> The first thing I did was go to the County Planning Office, where they were very helpful and gave me a lot of documents to read. But as I read, I became more troubled. One area of concern was dioxins, a fascinating chemical in many respects—especially for a chemist. . . . One of the first reports I read was from the California Air Resources Board from May of 1984 which said dioxin would be destroyed when heated above a certain temperature in the laboratory. . . . Would this also happen in an incinerator? . . . The document said, "Yes, based upon work reported by an engineer, Floyd Hasselriis." . . . He basically said that above a certain temperature there would be no dioxin.

Shortly thereafter Connett came across a challenge to Hasselriis by Barry Commoner and his group, who insisted that the original data did not support Hasselriis's conclusions. Connett went to the published sources of Hasselriis's report and found that they indeed did not support the claim that dioxins would be destroyed by high temperatures:

> So at this point I phoned up Barry Commoner—about in March of 1985. And to my amazement he picked up the phone, and I explained that I had just read his analysis of the Hasselriis graphs. . . . He said "If you're interested, next week at Hofstra University on Long Island both Hasselriis and I are down to give presentations, and I am going to challenge him in public to explain [the discrepancies]." . . . I went down. . . . Hasselriis produced the same graphs again, . . . and Commoner challenged him. . . . But Hasselriis had no real answer.
>
> You can see from the way I've dwelt upon this that it was a real

> turning point for me because I was appalled that he even attempted to do what he did with the [deficient] data he was using, and then appalled again that when it was made public all hell didn't break loose.

Connett was further disturbed by his local planning board officials, who seemed to take the whole issue of scientists fudging data for political purposes in stride:

> The local officials perceived me as an obstacle to building this incinerator, which they had to get around. As long as it was on the scientific level, they had to get around the science, but the moment it became obvious that the science they were relying upon was bogus—rather than admitting that, they used the high school debating tactic of saying things like "Well, I've met recyclers who lie." . . . In other words, "Whatever the problems you are going to throw up against incineration, there are problems with recycling too, so what's the big deal?" This is April of 1985, so then I knew what I was up against.

Returning to Dartmouth that summer, Connett became involved with another local incinerator conflict, where he brought the dioxin issue to the attention of citizens who were basing their challenge primarily on the noise level of the proposed plant. He used technical information from Denmark and Sweden on the dangers of dioxins. Connett noted that the health-risk assessment done for St. Lawrence County had considered only the danger of inhalation exposure to dioxin, whereas the Danish documents had emphasized that the danger of dioxins from the food chain was 500 times greater than that from the air, and the Swedes had put a moratorium on committing any more incinerators because of their concern about dioxins:

> [This dioxin data] was a shocker, and I thought with these Danish and Swedish documents that such information was going to stop the St. Lawrence County incinerator in its tracks because this is the largest milk producing county in the whole of New York State. . . . But I was going to be surprised because we had not won at all.

After giving a presentation in the summer of 1985 at Dartmouth, incorporating some of this information, Connett was invited to speak against other incineration projects in Vermont, New Hampshire, and Massachusetts.

> So now I'm in contact with Barry Commoner's group, who's fighting an incinerator in Brooklyn [New York]. We have people fighting the Claremont [New Hampshire] incinerator, people fighting the Rutland [Vermont] incinerator, and the people fighting the Holyoke [Massachusetts] incinerator during the first summer after I got involved.

Connett and Commoner joined forces to debate two industry representatives at the Vermont Law School in October 1985, and in the audience were people from different parts of New England:

> After the debate was over, I suggested that we all sit down, . . . and right there we formed the National Coalition Against Mass Burn Incineration and for Safe Alternatives. I volunteered to put out the first newsletter—it was forty pages long! . . . We charged five dollars for it. . . . The next newsletter took me nine months to get out because it was several hundred pages long and it came out in August 1986. . . . So the debate was in October, and it took till February [1986] to get the first newsletter out, and until the following August [1986] to get this big bugger out [showing a 200-plus page document]. . . . By August 1986, however, we've begun to network with many other groups. . . . This marked an important building of momentum for the anti-incineration movement.

Connett said that, until this time, incinerator proponents had been able to isolate and discredit any local opposition they encountered. It was common, he explained, for them to portray Barry Commoner as a radical who had run for president and had questionable scientific credentials:

> All of a sudden, the opposition was not isolated but rather networking around the country. And also, it was getting very well informed, and the worst thing for the incinerator industry is for their track record to be exposed. As long as they can come in with their PR presentations saying this is a brand-new incinerator which looks like a five-star hotel with a little spout on it—and you juxtapose that for

> local politicians with their awful landfill with the seagulls shitting everywhere. . . .
>
> It wasn't easy for an opponent before this to get through to Barry Commoner for assistance. . . . His reports were very technical, and there was little there which ordinary citizens could understand. . . . And just about this same time, the Citizens Clearinghouse (CCHW) also got involved in solid waste—because previously they had been primarily concerned with hazardous waste. . . . And they were very good.

In 1985, it was difficult for laypeople to get their hands on scientific materials that were useful in challenging WTE facilities. Ellen Connett emphasized the role of New York's Department of Environmental Conservation (DEC), which, she noted, actively promoted these incinerators:

> I believe that, in twenty or thirty years, what the DEC is doing today will be seen as major criminal acts. . . . They push incinerators down communities' throats, even though they give a lot of lip service to recycling. . . . The DEC goes along with the Mafia and organized crime which has run the waste industry in this state since the fifties. There's more money in incineration for them than in recycling.

Paul Connett brought the discussion back to the newly formed coalition's attempts to produce a regular newsletter responding to the need of people for more technical information about incineration and related issues, emphasizing how much he had learned from seeing another academic activist's more streamlined newsletter after being quite disappointed at his own first efforts in this regard:

> We realized that our first huge newsletter was ridiculous, that it was taking far too much of our time, and that it wasn't doing the job it was intended to do. We tried to get other member groups in the coalition we were forming to put this out, and somebody in Bloomington, Indiana, volunteered to do it, but couldn't. Then somebody in San Francisco got one issue out, got subscriptions, and never got No. 2 out because of family problems. . . . It became very sporadic so that, for example, when important news would come through I would be phoning every contact, etc. . . .
>
> In the meantime, we met Peter Montague and saw his "Rachel's

> Hazardous Waste News," and we got such a big kick out of it that we realized this was the way to do it: keep it short, keep it regular, send it with three holes in it with a number at the top so people could take the information, file it away. And what I really liked about it was that it was something you read when you received. Peter's newsletter was going for two or three years before this. It was our model. . . . Ellen decided to take on the responsibility, for pay of course, and she's been doing it since April of 1988. . . . We're now up to issue number 125, and I think it's one of the best things that we've done. We get compliments all the time, and I have to say it's my wife that does it.

In addition to this newsletter and telephone links around the nation, Connett also emphasized his national group's creation of twenty-three video tapes. ("I think it's over 3,000 copies now that have been distributed—and there have been some incinerators, such as one in Georgia, stopped on the strength of the videotape alone without our even having to go there.") There were also his own personal presentations across the United States, which had amounted to "over 600 in forty-two states in six years—tomorrow will be forty-three states when I go to New Mexico." Connett emphasized the scientific credibility that he saw his own work bringing to the WTE opposition movement:

> In September of 1985, this college [St. Lawrence University] financed my going to the Fifth International Symposium on Dioxin in Western Germany. That was the first time I'd been to such an international conference, and it was fascinating. If I had any doubt, it was firmly proven that dioxin was an extremely political issue, and it was very disturbing. . . .
>
> Risk assessment is full of politics. . . . Scientists who criticize the status quo must be prepared to be ostracized, but if there's anyone who gets it worse than scientists, it's farmers. . . . Farmers who complain are always told they are bad farmers, that they just do not know how to farm properly. I can give you chapter and verse on that. Go to Utah for the sheep that died from nuclear testing and the farmers were blamed. Go to Michigan where the cattle were poisoned and the farmers were blamed. Go to Bonny Bridge, Scotland, and the hazardous waste incinerator where the cattle died and the farmers were blamed. I was prepared because I saw what happened

to the scientists in the 1970s over the lead issue—how they were attacked in the scientific press. . . . We saw it with Barry Commoner up here [see Chapter 7], where Mary Verlaque who had absolutely miniature stature as far as anything scientific is concerned stands up and says that Barry Commoner is not a scientist. . . . So I was prepared for this. . . .

The thing with incineration is that when industry spokespersons say things like "Most scientists feel this way or that," they are only talking about scientists working for the industry. . . . Most other scientists don't know anything about it because they don't have any incentive to study the issue. . . . So our biggest problem is not being attacked by fellow scientists, but rather not having fellow scientists with enough time, energy, commitment or motivation to find out who the hell is right. You're out there by yourself, you're out there alone.

As far as the college is concerned, it's been wonderful. . . . The dean provided over a thousand dollars for me to go to that West German conference. . . . But I was more worried that people would view my spending enormous amounts of time on trash as not a very professional thing for a chemistry professor to do. . . . Yet the people in our chemistry department have been very supportive. . . . The industry has given me minor problems by calling up the college to try to get them to make damaging statements about me—two main issues they focused upon: "Is my work peer reviewed?" and "Does the college support me in my position?" . . . And the college has made it clear that it does not endorse my position as such, although I've had a couple of awards from the college which are its way of saying that they are not displeased with my activities. . . . To be real hard-nosed about this, St. Lawrence University sees that its name is getting out there many, many times. It sent me copies of recent clippings of my activities from newspapers all over the place during 1985 to 1988 which are eight or nine inches thick. . . . Occasionally, an editor will call me Professor NIMBY or something, but by and large these articles are very positive.

Connett emphasized how important the three "teams" that "keep me going" were. Besides his wife, Ellen, who has taken responsibility for their national newsletter, "Waste Not," he mentioned Roger Bailey, with whom he worked to make numerous videotapes, and Tom Webster, who worked for Barry Commoner. He met Webster in 1985, and "that's extremely significant because that's my scientific team. Tom and I have written practically everything together since that point."

> We have presented papers now at the last five international symposia on dioxin. I gave a paper in Japan in '86, Tom and I presented in Las Vegas in '87, and then in Sweden in '88, Toronto in '89, and we just presented our fifth. . . . The first four have been published, and the fifth is being submitted for publication. . . . There are a lot of questions about dioxins that are unresolved. It is very problematic. . . . There's a split in the scientific community, and what exacerbates that split is that there's a lot of people working for government agencies and they're not allowed to say too much, and people working for the incinerator and chemical industries who have an axe to grind. . . . It's only when they try to put one in your own backyard that most scientists become involved, and then they become what I call a NIMBI [Now I Must Become Involved]. . . .
>
> Most of the scientists we meet at these meetings who are not under pressure from anybody are concerned about dioxins and their impact on the environment and on human health. . . . A subset of these scientists are concerned about dioxins from incineration, and some of them are beguiled by the notion that with proper regulation such dioxins can be controlled. . . . The Europeans, especially the Germans, are much more sanguine about what technology can do than most are in this country, where we don't much trust our regulatory agencies because they are viewed as pawns of very strong economic interests. . . .
>
> We certainly don't trust the Mafia, and the Mafia's running the waste industry. . . . So what might apply in Europe doesn't apply in this country. . . . People talk about strong regulations, but they can only protect you where there is adequate monitoring and vigorous enforcement. Otherwise, the regulations mean beans. . . . There is what I would call the "survival of the incompetent" in regulatory agencies, the "survival of the boring." . . . I've said many times on public platforms that I do not think we are the victims of evil people out there, but rather of boring people, people with no imagination and no vision. They're the kind of people that survive. The others are either hired away or they have to fight like tigers to keep their

> jobs and funds in the face of state attempts to cut such activists' budgets or fire them.

Connett went on to list a few regulators whom he felt were exceptions at both the state and federal levels, people who brought to light evidence of toxic effects from hazardous wastes and similar pollutants, whose research grants and sometimes even their offices were taken away from them: Ward Stone, a New York wildlife pathologist; Gary Glass of the EPA; and Roy Gorman, a North Carolina toxicologist.

On his own involvement in opposing incineration, Connett observed:

> This work has taken over my whole life. I would say that in '85 it was within the bounds of reason, and '86 was when it really began to take off. Every time I got up and talked on this issue, somebody in the audience calls me back and says I heard you talk about the incinerator in such-and-such a place and would like you to come here. One thing leads to another.

It all started for Connett with his opposition to the incinerator in Ogdensburg, which was approximately seventeen miles northeast of his home in Canton within the same St. Lawrence County (see Chapter 7 for the details of this particular struggle). Connett explained that after he and an industry consultant disagreed about the effects of dioxin and cited different studies, an administrative law judge ruled it a technical debate that the judge himself was not qualified to referee:

> . . . The administrative law judge dismissed our case, taking the position that it was a scientific dispute in which he could not make a judgment about which of the two scientists was correct. And therefore because the Department of Health and the Department of Environmental Conservation said it was okay, and in the absence of any compelling evidence to the contrary, this judge decided to accept the Department of Health's word for it and ruled that therefore this dioxin is not an issue. . . . We had Austrian and Dutch scientists ready to testify in court, but we never got the opportunity to get the industry's experts under oath. That was a real disappointment for me. . . .
>
> We appealed his decision, but basically the judge would always await the pro-incineration lawyers' briefs and then use them in his

own. That's my interpretation of what he did. . . . This whole health-risk assessment is a bogus process here in New York State.

Reflecting on his experience in challenging incinerators in his own county and around the nation over six years, Connett discussed issue framing and gender concerns in grassroots mobilization over environmental justice:

> What we've learned from all this is that basically it is a political battle. If these incinerators get that far into your community, somebody's persuaded the politicians that this is a good idea. What you've gotta do is get into the system and persuade the politicians this is a bad idea. The politicians could be just naive and being misled, in which case you might get a chance to change their minds. On the other hand, these politicians may be being paid off and then you've gotta vote them out of office. . . .
>
> Environmental issues are what get the people emotionally involved because of their concern for their children, the threat of dioxin, and the heavy metals, so you shouldn't neglect the environmental aspects because it may keep some people fighting like tigers when others have given up. But in terms of winning a county, the real issue is usually economics. . . .
>
> The waste issue is the average person's most concrete connection with the global environmental crisis. And there is a tremendous army of people becoming involved, with the key soldiers in this army being women, particularly mothers. What they're doing is fighting for their kids, and they don't compromise when they're fighting for their kids.

Connett emphasized the independence among the local, state and national Work on Waste (WOW) groups he was instrumental in establishing:

> We're involved in both WOW St. Lawrence County and in the National Coalition Against Mass Burn Incineration and for Safe Alternatives. WOW St. Lawrence County was a member of the national group. . . . Most of the people in WOW, however, have not become very much involved in the national issues. . . . Ellen and I have been very much involved in networking at the national level. . . . We changed the name of our national organization to Work on Waste USA because neither we nor others could remember the other name very well. . . . After that we formed Work on Waste New York State,

> which is a statewide coalition. . . . There was a point where Work on Waste USA could have become a very structured organization, but we decided against that because we wanted to keep it as a network where you don't control anybody, you do what you want—if you're against incineration, you're one of us.

Although outspokenly unimpressed with his adversaries' abilities to separate their science from politics, Connett emphasized his own accomplishments in this regard:

> Industry didn't realize that I could keep the science and the politics separate. . . . You can look at our scientific papers until you're blue in the face, but you won't find any politics. . . . Tom Webster is a brilliant mathematician. . . . They have been unable to dismiss our science. . . . If you like, of course, you can say that the reason we did the science is because we were politically involved with the incineration issue. . . . I don't lie, and can't be bought—and that's another thing they find it very difficult to deal with. . . . I do take contributions for my work, but nothing like those bozos get on the other side.

Connett also reflected on the similarities he perceived between the incinerator and nuclear power industries:

> As people get more and more concerned about air emissions, and the technology gets increasingly expensive, the incinerator industry is really going down the tubes, just like nuclear power. . . . What's happened in Europe is very important because they're talking in Germany about standards for a medium-sized incinerator, which will cost between 25 and 30 million dollars to retrofit to meet the new dioxin requirements. . . . They're retrofitting incinerators in Holland, Germany and Austria right now with activated charcoal filters. . . . So incineration is going to get more and more expensive, just like nuclear power before it.

He went on to discuss some of the companies involved in designing and building modern incinerators:

> Although all of the leading companies in the [incinerator] industry are not former nuclear power plant builders, some such as Westing-

> house, General Electric, Babcock and Wilcox, and Combustion Engineering certainly were. . . . Another way to put it is that they are all major engineering firms, and major engineering firms have got to build something. Incineration was seen as the biggest boost to construction engineering since nuclear power.

Connett concluded by suggesting that just as Europe had preceded the United States in getting into modern incinerators, we might look there and find even the most previously enthusiastic supporters turning away from this technology because of concerns with such things as dioxin, mercury, and toxic ash. He admitted being very disappointed with the published report by the U.S. Congress Office of Technology Assessment—even though he was himself on its Advisory Panel—and he criticized the way industry can prepare ahead of time when it knows twenty-four days or so in advance the date on which dioxins are going to be measured. Despite what he regarded as a regulatory stacked deck in the industry's favor, however, Connett felt confident that dioxin and ash problems would eventually overwhelm it:

> Europeans are very much more concerned about dioxin than Americans—dioxin in the food chains and dioxin in mothers' breast milk. Although incinerators are not the only sources of dioxins, they are certainly a significant one. . . .
>
> The better they get at protecting the air, the worse the ash is going to get. . . . The worst thing you can do to toxic metals is to burn them. . . . Take this bit of plastic here [picking up a small container], and suppose it has a cadmium additive to stabilize the plastic. There's no way I can get the cadmium from the plastic into my nose and into my lungs. If it gets buried into a landfill, it may take as long as a thousand years for that cadmium to leach out of that plastic because it's a very stable matrix. However, if I burn it, that cadmium will come out either as a gas or as a tiny particle which can go deep into my lungs. So the worst thing you can do with things that have toxic metals like mercury, cadmium, chromium, arsenic, and lead is to burn them. . . . And trash incineration is the second largest source of mercury in the United States, after coal burning. . . .
>
> In Bavaria, the home of trash incineration in Europe, they've built seventeen incinerators in twenty years with fifteen more planned. . . . They've ninety groups organized against incinerators there because

> doctors are concerned about respiratory problems, and because of the build-up of dioxins in the environment and in mothers' breast milk. . . . Bavarian doctors are much more concerned about dioxin in mothers' milk than doctors in this country. . . . So, one of the most pro-incinerator provinces in Germany which is itself one of the most pro-incinerator nations is likely to say "no more incinerators in Bavaria." . . . The U.S. builder Ogden-Martin is based in Bavaria, and has long used as one of its selling points that if there were anything wrong with this technology it would have been found out long ago in Germany.

Chastened Industry Perspectives

By the late 1980s, incinerator proponents might disagree with such dismal assessments about the future of the technology, but they had to acknowledge the significant impacts of Connett and the organized protests of the mid-1980s on incinerator siting processes. It is instructive to notice the strong emphasis in the industry's own journals by the late 1980s on incineration as merely a single component of a more comprehensive system for handling municipal solid waste.

For example, a Westinghouse official writing near the end of the 1980s told his pro-incineration readers they had to change their attitudes if they wanted to continue in business. Cautioning against their tendency to view the WTE plants themselves as major solutions to the nation's trash problems, he insisted that "waste management and disposal is much broader than designing, building, and operating waste-to-energy plants":

> We are not in the waste-to-energy business. We are in the business of providing a variety of environmentally sound solutions to solid waste problems. . . . It is really a business with four essential parts: waste reduction, recycling, resource recovery, and landfills. . . . Resource recovery plants do not offer a single solution to waste disposal. And those of us who think of ourselves merely as waste-to-energy plant vendors and operators may not survive in the 90s. (Pollier, 1989:6)

This writer and other industry consultants were now echoing the priorities emphasized earlier by West Europeans and Japanese: the reduction of as much MSW as possible at the source, maximum recycling and/or composting of whatever MSW is generated, incineration of whatever MSW cannot be recycled or composted, and landfilling of the rest:

> In America, with our vast geographical expanse, few people thought there would ever be any limits on our ability to keep finding new places where we could dump our garbage. For far too long, municipal solid waste officials felt their main job was to find out-of-the-way and out-of-sight locations to landfill waste. This strategy worked for a long while.
>
> But it doesn't work anymore. It doesn't work when groundwater moves pollutants into water supplies; it doesn't work when communities are banding together to prevent landfills from being sited near them; and it doesn't work when cities and states are sealing their borders against the importation of waste from other jurisdictions. . . .
>
> The 90s will demand a broader view—including the ability of potential [incinerator] vendors to meet existing and emerging recycling demands and environmental criteria. . . . Events of the last decade have ushered in a new age of environmental awareness that will be with us for the rest of this century. The images are striking: Bhopal, Chernobyl, the greenhouse effect, toxic waste, garbage and ash-laden barges looking for homes. All of this has helped change the way most of us look at the environment. Now we see an environmentalist every time we look in the mirror. (Pollier, 1989:5–6)

A writer in another industry journal acknowledged that the anti-incineration perspectives common by the late 1980s had created a political environment which could be characterized as adding "uncertainty to even the most technologically and economically sound projects" (Hocker, 1991:12).

An additional issue that emerged in the course of the conflict between proponents and opponents of modern incinerators was "flow control," which will be central in one of the New York disputes analyzed in Chapter 7. As the phrase suggests, this refers to the right of county officials to control the flow of trash to a particular facility. The key question is whether authorities should have the right to order all of the municipal waste from any particular area to a particular disposal facility, or whether trash haulers should be allowed make their own decisions based upon cost and other factors. An industry manager looking back upon the conflict from a 1993 perspective emphasized the importance of flow control, as well as air, ash, recycling, and what he considered biased media coverage in the political struggle over incinerator siting:

> In the early 1980s, financial analysts and industry executives were awaiting a predicted boom in WTE facility construction. Waste generation levels were expected to increase at a fast pace, and landfill capacity was decreasing. . . . Financial firms, law firms, design/construct firms, consultants, and full-service vendors all positioned themselves to capture a piece of this new market; the number of major vendors proposing to build and operate facilities exceeded a dozen. A decade later, however, the boom has yet to occur.
>
> The 1980s were not a complete bust for WTE. More than 100 facilities were constructed and there are now 143 facilities in operation. . . . Despite the occasional "successes" for WTE, the fact remains that WTE has fallen short of the expectations of industry professionals. . . .
>
> What proponents of WTE did not anticipate was the fervor of the opposition driven by environmental concerns and, as a result, the lengthy—and, in some cases, aborted—siting and public acceptance process. . . .
>
> Over the years, opponents of WTE projects have used a variety of environmental issues to delay or prevent the implementation of projects. One such issue is mercury emissions from WTE facilities. . . . The evidence compiled to date shows that mercury emissions from WTE facilities are insignificant in comparison to the total amount of mercury emitted to the environment. However, as a result of the debate . . . all of the WTE facilities recently receiving permission to construct include add-on mercury controls to reduce the emissions of mercury. These controls will be installed because, without clear guidance from the regulatory community, project proponents must install all available technologies to mitigate public opposition and avoid further project delays. . . .
>
> An element of organized opposition that often frustrates engineers and scientists is the lack of accountability for statements made to the press or before elected officials. Misstatements of facts often find their way into print and achieve, thereby, an undeserved mantle of credibility that can influence decisions of elected officials regarding the direction of MSW management for their jurisdiction. The issue of "toxic ash" is an example.

Although ash residue from WTE facilities contains organic compounds and heavy metals that are of concern if released into the environment, there is no evidence that these compounds cannot be properly managed and contained when properly landfilled. In fact, considerable scientific evidence shows that if managed correctly, the disposal of ash potentially poses less risk to the environment than landfilling raw MSW.

The ash residue disposal issue continues because certain laws are inconsistent with the scientific evidence. While a number of states and the EPA have addressed the issue and regulate WTE ash as a special waste, Congress continues to debate whether ash residue should be managed as a hazardous waste or special waste. Without resolution, the public is left to decide for itself and support a cautious approach, thereby substantially increasing costs for the development of a WTE facility. . . . Experience has shown that in properly designed integrated waste management programs, WTE and source reduction and recycling programs are not only compatible but complementary.

Flow control—one of the pillars of most WTE facility financing—. . . is now being challenged as unconstitutional. Flow control typically refers to local ordinances or state enabling statutes that stipulate that all waste generated within a defined geographic area must be disposed of at a location specified by the local government. Although the concept of flow control is quite old, it was not used extensively until the development of highly capitalized WTE facilities. Most WTE projects are financed based on revenues generated from the receipt of waste. Flow control ordinances are enacted to ensure that the waste generated in the planning area for which the disposal facility is constructed is actually delivered to the facility.

In most states, local government has the responsibility to provide for the safe and adequate disposal of MSW generated within its borders. . . . Regardless of whether an existing or new facility is stipulated in the waste management plan, the control of the waste for which the plan is generated is often essential to the effective implementation of the plan. . . . During the early development of WTE facilities, the constitutionality of flow control was addressed a number of times and the courts upheld the right of local government to enforce flow control ordinances. In the last few years the opposite has occurred—some lower courts have ruled against local governments and struck down the constitutionality of flow control. . . .

Another type of flow control currently being debated is the importation of waste. Contrary to early court decisions regarding the export of waste, courts in past years have found that banning the importation of waste by a state was unconstitutional. . . .

Having addressed many technical roadblocks over the years, WTE now faces institutional road blocks as well, which engineers and scientists will have considerable difficulty overcoming. (Demme, 1993:92–96)

Does WTE Have a Future?

While opponents such as Paul Connett insist that the death knell has sounded for the WTE industry, many incineration proponents disagree. Although the latter acknowledge the effectiveness of the anti-incineration movement in dramatically slowing the industry's growth during the 1980s, and complain that the federal Clean Air Act amendments of 1990 as well as the EPA's Emission Guidelines of 1991 will mean expensive retrofits of pollution control systems, they foresee a resurgence of incinerator sitings. New companies have become involved in building these plants, and they expect opportunities to expand in the near future. For example, the vice-president of one new WTE firm said: "[By the late 1990s] we think the public will be more educated about the benefits of WTE, and public agencies will have developed integrated waste management programs that include incineration along with source reduction, recycling, and landfills" (Hocker, 1991:13).

Industry executives also suggest that public attitudes toward recycling will become more "realistic" because "after you achieve the maximum in materials recovery and recycling, you're still left with about 75 percent of the original waste stream for incineration or landfilling" (Hocker, 1991:17). Some environmentalists, on the other hand, insist that solid waste management officials who assume a 25 percent recycling rate maximum are likely to actively subvert efforts to achieve even higher levels of recycling if they perceive them as jeopardizing the amounts of trash available for their projects.

Whatever the future for WTE facilities, opposition has been mobilized and the era of routine sitings is over. The three states in which our projects were located—New York, New Jersey, and Pennsylvania—are nationally ranked first, fourth, and sixth, respectively, in terms of their total numbers of WTE plants (Hocker, 1991:16). Each of the siting attempts that are the focus of this book was in process after the mid-1980s when the anti-incineration movement began to hit its stride. Some were defeated by organized citizen opposition, while others were completed, and our task is to account for these divergent outcomes.

2

Theoretical Perspectives on Technology Protests

Both sides, in each of the challenges to modern incinerator projects that are analyzed in this book, claimed to be operating in the public's best interests. While we try to be evenhanded in analyzing these conflicts, our primary focus is on the grassroots protests as viewed through the lenses of contemporary social movement theory and environmental sociology.[1] A selective summary of the evolution of social movement theory in recent decades provides a useful background for understanding our theoretical perspectives.

What Is a Social Movement?

Movements of various political stripes have become increasingly common in the United States in recent decades. African Americans, Native Americans,

1. For more of an industry perspective on the same general topic, see Curlee et al., 1994, a book that came to our attention as we were completing this one. Although it focuses primarily upon aggregate social and economic indicators, rather than on the actual dynamics of specific grassroots struggles, it is interesting to see some surprising similarities in conclusions about processes and outcomes arrived at via their "top down" and

women, farmworkers, environmentalists, consumer groups, and scores of other collectivities have been more or less successful over the past few decades in mobilizing to make demands for political and social change. The 1960s were particularly turbulent times, with their numerous and overlapping movements, such as the civil rights, students, anti–Vietnam War, women's, and gay and lesbian movements, and even the beginnings of an environmental movement. In comparison, the 1970s and 1980s seemed relatively calm, but they also had their own forms of collective protest. Organized anti–Vietnam War activity continued into the early 1970s and is commonly acknowledged to have shortened the direct military involvement of the United States in Indochina. The citizen mobilization efforts against nuclear power—another example of a social movement that came of age during the 1970s—also became a serious threat to that particular industry. Widespread citizen protest against U.S. involvement in Central America, and mobilizations on behalf of animal rights, are examples of organized protest that emerged during the 1980s. In addition, there have been numerous local protests against technological threats, such as oil spills, chemical plants, hazardous waste sites, overflowing landfills, waste-to-energy incinerators, and other perceived hazards to citizens' health and safety. Such social movements—at national, regional, and/or local levels—have thus become increasingly common and important political instruments for citizens working toward social change in the 1990s.

While some view social movements as tools of radical or liberal groups in pursuit of their own ends, the actual evidence reveals a variety of such phenomena across the political spectrum. Historically, many have been situated toward the political left because grassroots mobilization is typically an instrument of those who have little formal political power. Yet there have also been movements directed primarily toward personal transformation, and today we see an increasing number of contemporary movements championing familial structures and ethical perspectives they consider endangered. The Moral Majority, for example—which opponents' bumper stickers insisted was neither—was perceived by some of its critics as having taken over the reins of the federal government through the Reagan administration in 1980, and the Religious Right continued to exert significant political influence, especially in the Republican Party in the 1990s. Because

---

our "bottom up" approaches to these incinerator siting conflicts. Because they virtually ignore the national anti-incineration movement and its impact on siting processes since 1985, however, these authors fail to consider numerous important variables influencing outcomes.

conservatives are by definition more aligned with the established order than reformers or radicals, however, many expect them to favor routine electoral processes rather than social movement strategies to accomplish their goals. While analysts still debate the precise nature of the causes for this shift, movements of the "Right" became increasingly common after the early 1980s. In the wake of the 1995 Oklahoma City bombing, for example, the media began concentrating attention on what some refer to as the "Militia Movement," which has an ambivalent relationship with other conservative groups.

Because the term "social movement" has been applied to a wide variety of phenomena over the years, controversies about its meaning abound. When revolutionary uprisings like those responsible for changing the governments of the United States in 1776, Russia in 1917, and China in 1949 are called "social movements," and the same term is also applied, for example, to such relatively low-key mobilization efforts as the transcendental meditation (TM) or Mothers Against Drunk Driving (MADD) movements, we are unlikely to discover many useful general patterns across such processes. Recognizing these problems, certain writers have suggested that the "social movement" tag should be reserved only for organized *political* protests. Others prefer to use the concept "social movement" more broadly to include, in addition to political protests, organized collective action focused on religion and self-change—insisting that any distinction between religious and political movements is more semantic than real, because there is enough dogma in politics, and political power in organized religion, that any rigid distinction between the two realms is problematic. Rather than attempting to turn back the fog by challenging these popular, vague, and broad uses of the "social movement" label—or even trying to restrict its usage to one or another genre of mobilization—we prefer to settle for a broad definition, such as: an enduring challenge to established elites by an organized collectivity with a common ideology. Within such an inclusive delineation that roughly corresponds to common usage, analysts are free to specify subtypes of movements revealing similar patterns—distinguishing political revolutions from ethnic mobilizations and religious conflicts, and both from more particularized technology protests, like those considered in the following chapters.

One example of this strategy is the label "new social movements" (NSMs), which certain West European analysts apply to their own national strands of the environmental, women's, and peace movements to distinguish them from the earlier workers' movement, which they consider the archetype of an "old social movement." These theorists emphasize what they consider major differences between "new" and "old" movements, insisting that the new social movements are distinct in terms of their social support bases, goals, structures, and styles. For a variety of reasons, however, few U.S. analysts find this contrast between "old" and "new" social movements useful, primarily because of societal differences between Western Europe and the United States in protest experience. Two of the most important specific discrepancies between them are that the labor struggles of the 1930s were less dramatic in the United States and that the continuities, rather than the differences, between the numerous social movements emerging in the United States since that time have usually been emphasized. Two major U.S. social movements of the 1960s—the civil rights and anti–Vietnam War movements—did not really have counterparts in Western Europe, and the lack of such major bridging mobilizations between the workers' protests of the 1930s and the contemporary new social movements made that difference all the more dramatic for European analysts. In other words, societal variation in protest experience helps explain why the NSM label is more commonly used in Europe than in the United States (see Klandermans, 1986; Walsh, 1988a).

Are Technology Movements Different?

Increasingly, U.S. analysts are now calling attention to a novel genre of protest movement that has distinctive characteristics but no widely accepted label as yet. One researcher, for example, alludes to "a whole new species of trouble" in the guise of invisible threats associated with toxic poisons from modern technologies (Erikson, 1991). Different tags have been given the broader movement enveloping these hundreds of loosely networked local protests. Some refer to them as the grassroots component of a two-tiered environmental movement, where the upper class is represented by such national lobbying organizations as the Sierra Club and the National Audubon Society (Cable and Cable, 1995:67). Others prefer the designation "toxics movement," suggesting that it may be only the beginning of something evolving into an eventual reinvigoration of progressive politics in the United States (Szasz, 1994). Still others use "environmental justice movement" to refer to the complex mix of factors involved in grassroots protests against technologies involving invisible environmental poisons detectable only with sophisticated scientific equipment.[2] We have previously used the label "technology movement" to analyze comparable protests (Walsh, 1988b; Walsh et al., 1993). Because this novel variant—whatever name it bears—is similar in numerous respects to other movements, we summarize the commonalities before noting the idiosyncrasies.

Social movements oriented toward political change are typically based on discontent stemming from conflicts of interest inherent in established institutions. Widespread grievances, caused by perceived abuses of official power, often give rise to a collective struggle led by one or more protest organizations against incumbent authorities. While some analysts emphasize the importance of lead organizations in generating the necessary discontent, others insist that people's grievances contribute their own momentum to the formation of protest organizations. Sometimes, widespread discontent precedes the attempt to organize—think of the civil rights movement of the 1960s by millions of African Americans and their supporters as a response to centuries of injustice and discrimination. The women's movement is another example of mobilization following widespread and protracted grievances. At other times, however, organizations are more obvious generators of discontent—for example, the efforts by various organizations to create and simultaneously mobilize public opposition to U.S. intervention in Central America during the 1980s. Frequently, there is a reciprocal relationship between discontent and organization whereby they feed on one another in the mobilization of collective protest, so that a vague preexisting discontent is focused for political action by emerging protest organizations, as in the civil rights and women's movements. It may also happen, however, that small existing protest organizations are unable to mobilize widespread support from the public until the latter are confronted with the very grievances predicted by the initial activists—as happened in the anti-nuclear movement after the Three Mile Island accident.

The kinds of questions researchers studying social movements pose have changed considerably over recent decades. Before the 1960s, for example, it was common to focus on such issues as "What kinds of people join social

2. Disagreement among academics and activists about the appropriateness of labels is common. An example of such differences of opinion was given by Robert Bullard, who told of his own public disagreement with leaders at the Citizens Clearinghouse for Hazardous Wastes over the use of the "environmental justice" label, which he preferred to see limited to protests where ethnic minorities figure prominently in the dispute. Bullard made these comments at a Georgia Institute of Technology consortium on negotiation and conflict resolution in November 1994.

movements?" or "Why do these people want social change so much?" Part of the explanation for this focus was that major movements of the era were led by such organizations as the Communist Party or the Nazis, neither of which was favored by most academics doing the social movement research. It was hardly surprising, therefore, that they came up with unflattering characterizations of movement participants and their motives. During the 1960s, the emergence of new movements more favored by those doing the research prompted a shift of perspective and new types of questions. Studies during that time comparing participants and nonparticipants in such movements as the civil rights, women's, antiwar, and student protests turned up no negative psychological data on participants. To the contrary, the latter were commonly said to be more socially integrated and even, in some studies, more psychologically balanced than nonparticipants (Marx and Wood, 1975).

By the early 1970s, increasing numbers of social movement researchers had abandoned the assumption that participants in social movements were themselves deviant, and they were attempting instead to answer some variant of the general question "Which factors promote and/or retard successful protest mobilization?" The focus had moved from a preoccupation with psychological questions about movement participants, previously presumed to be somewhat unbalanced because they were dissatisfied with existing societal arrangements, to more organizational and structural questions about the effectiveness of mobilization processes and the outcomes of social movements. This new approach came to be referred to as "resource mobilization" (RM), and it had become dominant by the early 1980s. Rather than asking psychological questions about the individuals who joined this or that movement, researchers were increasingly focusing attention on various aspects of the organizational carriers of protests, commonly referred to as social movement organizations (SMOs).

Some of the pioneers of the resource mobilization approach went so far in their emphasis on organizational over psychological factors for understanding mobilization processes that they were even willing "to assume there is always enough discontent in any society to supply the grass-roots support for a movement if the movement is effectively organized and has at its disposal the power and resources of some established group" (McCarthy and Zald, 1977:1215). Questioning this organizational emphasis, other researchers—in general sympathy with the new RM perspective, but concerned about its tendencies toward structural reductionism—insisted on including such mental factors as grievances, ideologies, and framing processes, in addition to organizational factors, for a better grasp of the complexities involved in actual protest mobilization processes (Useem, 1980; Walsh and Warland, 1983; Snow, Rochford, et al., 1986).

Contemporary social movement perspectives emphasize the complex structural as well as social psychological interactions among movement activists, their protest organizations, and the institutional environment. Social movements emerge and evolve within the context of existing society. For instance, the media usually have a more ambiguous role vis-à-vis protest movements than some industry that is under direct challenge from organized opponents—as is the case, for example, with the modern incinerator industry. The media need the news generated by protest organizations—just as such protest movements need the exposure provided by the media. One analyst suggests that this dialectical bond between the media and movements might best be described as a "dance—sometimes the dance of death" (Molotch, 1979:92). The resource mobilization emphasis on social structural analysis also prompts the recognition that the media itself is increasingly dominated by economic elites.

According to resource mobilization perspectives, then, challengers attempt to mobilize available resources and create new ones in pursuit of their goals, while authorities opposing them may be expected to manipulate the existing system in their efforts to undermine such protests, both directly and indirectly. The use of a police force to disperse a grassroots rally, for example, is a direct repression tactic, while the threat of a lawsuit, or even the subtle suggestion that activists may jeopardize their occupational careers by participating in collective action, illustrates indirect social control. Whereas some RM analysts emphasize the central importance for grassroots protests of outside resources from third parties not originally involved in the conflict, others insist that the aggrieved often have their own "preexisting institutional and cultural skeleton of opposition" (Morris, 1992:371) that tends to be ignored when too much emphasis is given the role of outside resources. In the civil rights movement, for example, there was indeed outside assistance from the liberal establishment in the North, but also internal support from the ideas and networks of African Americans' own Southern churches, labor unions, voluntary associations, music, informal conversations, humor, and collective memories of elders participating in earlier struggles. Another way of phrasing this critical issue is to ask about the relative importance of indigenous vis-à-vis outside resources in effective protest mobilization. We shall return to our technology movement variant of this distinction when we compare the relative importance of indigenous backyarders' resources and those of activists outside the backyard areas in the various counties where these incinerator projects were being considered.

Early environmental challenges were virtually ignored by social movement analysts because of the widespread perception that they were atypical of grassroots protests. Some early critics, for example, emphasized elitist aspects of the environmental movement of the 1960s and 1970s (Morrison, 1986), but it had been so transformed by involvement in grassroots struggles during the 1980s that later analysts emphasized the central importance of working-class organizations challenging environmental degradation (Cable and Cable, 1995), and others even suggested that it might become the unifying element in a contemporary mass movement for social justice (Szasz, 1994). Recent challenges to incinerator sitings based on complaints about air pollution and toxic ash are part of this new grassroots insurgency. If such protests have typically been based in working-class neighborhoods, they have also included large numbers of middle-class participants. As Paul Connett put it: "Trash problems are most people's most tangible link to environmental issues."[3] Contemporary environmental activism, however, forces analysts to think of the broader movement in terms of a continuum ranging from mainstream political organizations and lobbyists—its "upper class," so to speak—to more radical neighborhood social movement organizations situated within or adjacent to areas zoned "industrial." Whatever their earlier accuracy, charges of the environmental movement's "elitism" became obsolete when less affluent local protest organizations began playing such a major role (Buttel, 1987).

Because this multiplication of grassroots protest groups during the 1980s represented a fundamental change, it forced national environmental organizations to reevaluate their ideologies, strategies, and tactics (McCloskey, 1992). In addition to incorporating a wide variety of new issues, such pressures also forced the larger environmental movement to become increasingly global and critical in perspective (Caldwell, 1992). At the interface of environmental and more mainstream social movement issues is the claim that locally undesirable land uses (LULUs) are much more likely to be attempted and carried out in minority communities—or, at the global level, in more vulnerable nations. In challenging such attempts, the struggle for environmental justice is viewed by some as indistinguishable from a civil rights protest. African American and other communities of color have been

3. Connett interview.

documented as bearing disproportionate waste disposal burdens in the southern United States (Bullard and Wright, 1992).

This increasingly radicalized environmental movement played a central role in the community protests discussed in subsequent chapters. As the range of problems the movement addressed expanded from hazardous waste dumps to incinerator sitings, for example, its organizational base also swelled with increasing support from threatened backyard communities, women, and minorities (Dunlap and Mertig, 1992). Important links between the broader environmental movement and the particular protests analyzed in this research—such as Barry Commoner's influence and of Love Canal's Lois Gibbs—were mentioned in the previous chapter.

To narrow the often-lamented gap between theory and data in the social movement literature (Zald, 1992), we reexamine some widely accepted generalizations in the light of our empirical findings involving grassroots environmental disputes. The evidence reveals that each of these protests is really part of a loosely linked national movement that, especially via its grassroots components, illustrates the crossfertilization of environmental sociology and social movements currently in process (see Cable and Cable, 1995; Schnaiberg and Gould, 1994; Szasz, 1994). Grassroots challenges are the cutting edge of the broader environmental movement, and in the following chapters we identify key factors promoting and militating against their success. In the course of our analyses, we address such "underdeveloped areas" (McAdam et al., 1988) of social movement research as protest development over time and decision-making at the organizational level. No general explanatory model fits all types of social movements, but distinctions between equity and technology movements are theoretically useful.

In our initial efforts to identify the variables promoting and militating against incinerator sitings, we consulted both the social science literature and players on each side who were acquainted with the specific processes. The informed observers provided many useful insights, but they also suggested that each project was essentially unique because idiosyncratic factors figured prominently in sitings as well as defeats, and that made generalizations about outcomes virtually impossible. Socialized as we have been to believe that closer study commonly reveals latent patterns, we interpreted their assessments as a challenge. Because each case included comparable grievances and clearcut outcomes, we assumed that careful analysis would reveal configurations distinguishing successful sitings from defeated ones. Realizing that activists at each site framed their discontents somewhat differently, we felt that their common opposition to similar incinerator projects would make comparisons between and among them instructive. Because some of the proposed facilities were constructed despite such opposition, while others were defeated, we also had the opportunity to consider multiple examples of similar protest phenomena where the outcome indicator (success/failure) was relatively unproblematic.[4]

Central Concepts in This Study

By the mid-1980s it was the exceptional waste-to-energy project that did not meet at least some organized citizen opposition from the proposed host community, often supplemented with outside assistance from the emerging national movement. The existing literature suggested certain characteristics of grassroots groups likely to be associated with protest mobilization. Environmental justice researchers, for example, predicted disproportionate percentages of minority group members in backyard communities surrounding such projects (Bullard and Wright, 1992). Resource mobilization theorists suggested that the higher the target community's socioeconomic status, the greater its resources for challenging such projects were likely to be (McCarthy and Zald, 1977). The greater the intensity of grievances in the same backyard community, the more likely such protests were to emerge via either existing or newly created protest organizations (Useem, 1980; Cable et al., 1988). The more dense the existing friendship and associational networks among the people in these target communities, the more likely grassroots opponents would be to mobilize successfully against such projects (Tilly, 1978). Contemporary social movement theory also predicted that backyarders who were not forced to contend with organized pro-incinerator forces, referred to in the literature as countermovement organizations (Zald and Useem, 1987), should be at an advantage over comparable communities where such countermovement organizations worked to neutralize protesters' efforts. Such factors are commonly said to facilitate mobilizing efforts for all types of movements, ranging from the labor struggles of the 1930s to alternative religions in the 1990s. While these and related insights from the literature were useful in our research design, we also expected to find

4. For an example of the types of problems that researchers who are dealing with more problematic measures of social movement "success" typically confront, the Goldstone-Gamson exchanges in the *American Journal of Sociology*, March and May 1980, are instructive.

differences between the "equity" movements serving as primary empirical referents for contemporary social movement theory over the past few decades and these more recent examples of "technology" protests.

Equity and Technology Movements

Although organized protests emerging in response to suddenly imposed environmental grievances reveal numerous similarities to other types of collective action, their peculiarities also warrant special attention. The ideal type of what we refer to as an "equity movement" involves a gradual mobilization around long-standing grievances among members of a culturally identifiable collectivity—for example, African American, Native American, or women—which depends for its success on the appropriate combination of political opportunities and internal organization among those involved.[5] The typical technology movement, on the other hand, emerges more quickly in response to a perceived environmental threat that spurs mobilization of local residents within some geographical area against the industry in question (see Morrison, 1986; Walsh, 1988b). Equity movements usually involve "soft" grievances—grounded in demands for equal rights—that have festered for decades among deprived people. Technology movements typically focus upon "hard" grievances, such as a proposed hazardous waste facility, nuclear power plant, or municipal incinerator in some specific geographical area, and are at least partially grounded on scientific arguments regarding the effects of invisible radiation, toxic chemicals, or equivalent threats to health and safety.

While equity and technology movements are concerned with both economic and political injustices experienced by underdogs, for example, technology movements usually involve essentially invisible radiation emergencies or the threat of toxic poisons. Citizen protests against power plants, waste dumps, incinerators, and similar targets are increasing in the United States and around the world in response to such threats. Whereas equity movements seldom contend with mainstream spokespersons arguing

5. A thorough discussion of equity movement models is McAdam, 1982, where the author contrasts what he defines as a "deficient alternative" model (which tends to ignore subjective variations in discontent, and especially the role of a movement's mass base in the generation of insurgency) with a "political process" model emphasizing the importance of the interaction of expanding political opportunities, the indigenous organizational strength of the target collectivity, the presence of shared cognitions within the same target population, and shifts in the response of outside groups to the movement. A more recent synthesis of literature on social movements and revolutions is Tarrow, 1994.

against them and in overt support of the "soft" grievances (for example, racism or sexism) being challenged, technology protests typically are confronted with industry spokespersons and must depend for scientific credibility on maverick scientists to make credible responses. And although we have here presented equity and technology characteristics as clustering together into two contrasting types, others may want to treat the correlations between and among dimensions as research hypotheses for additional study (see Snow, Cress, et al., 1996).

Success and Displacement Goals

Determining "success" for any social movement is problematic, but the tangible goals of technology movements make the researcher's task more straightforward. On the other hand, when analysts attempt to measure the predictors of success for equity movements, the criteria and time frames for the dependent variable are open to debate. Their only clear correlate of failure is seeking to eliminate or displace antagonists rather than coming to some kind of compromise agreement, and even here there is confusion over the operational definitions of success (Goldstone, 1980a). Organized protests against the proposed restarting of a nuclear power plant or construction of a hazardous waste facility are more conveniently judged a success or failure: the nuclear facility is either shut down or remains on line; the hazardous waste dump is sited or rejected. Such struggles are also especially noteworthy because they necessarily involve the allegedly risky goal of attempting to displace, at the local level, the protest movement's antagonist. The mobilization attempts analyzed in this book provide an opportunity to examine the dynamics and correlates of successful local protests where such displacement of their antagonists is the primary goal of grassroots activists.

Even after effectively mobilizing, technology movements confront unique challenges in attaining "success," because if they turn to the courts they are often forced to argue their cases before federal or state regulators with symbiotic links to the industry in question, or before judges who are intimidated by the arcane nature of the technology at issue. As a former commissioner of the Nuclear Regulatory Commission pointed out, the courts commonly defer to the industry over the regulators when challenges from these parties concerning high technology issues come before them, and to the regulators over citizens when these are the disputants. Such a pecking order, with the industry on top and citizens on the bottom, does not bode well for grassroots legal challenges (Bradford, 1979).

Networks

Social movement analysts assume that protest mobilization varies directly with the density of preexisting organizational and friendship networks among opponents (Tilly, 1978), as well as with the amount of organizing activity previously carried out among the collectivity in question (Zald, 1992). The obvious implication here is that the roots of activist recruitment are embedded in structural networks—of family, friendship, neighborhood, church affiliation, the workplace, voluntary associations, and the like—rather than in individual psychological dispositions. Thus, the successful mobilization of a protest movement, according to these theorists, depends upon the strength of preexisting networks among its potential supporters (Knoke, 1990). A few analysts, however, have questioned what they regard as an exaggerated emphasis on existing networks to the exclusion of networks newly formed as a result of a suddenly imposed grievance or moral shock (Jasper and Poulsen, 1995; Cable et al., 1988).

Networks are said to provide the building blocks for the broader protest structures that eventually emerge in successful grassroots organizing efforts. The latter have been compared to the "hardware" that then must "create and apply a kind of 'software,' " or framing of the issues, to attract public support (Gerhards and Rucht, 1992). Thus even in attempts to integrate social psychological principles with RM theory, network elements figure prominently in explaining why people want to participate in social movements (Klandermans, 1984). We consider the importance of networks, both within and outside the backyard areas, in helping to explain which of the projects we examine were defeated and which were sited.

Framing Grievances and Ideology

In their efforts to emphasize what they regarded as important structural factors, some theorists minimized or completely ignored grievances and ideology that influence social movement emergence and development. Critics subsequently pointed out both theoretical and empirical reasons for including grievances and other mental factors in such accounts, and it has now become standard practice to refer to this whole cluster of social psychological and signifying concepts with the notion of "framing."

Framing refers to the reality construction processes by which any challenging group decides upon a public presentation of its grievances, perspectives on problems, and possible solutions. "Frame alignment" is a concept suggesting the linkage of individual and SMO interpretive orientations, so that some set of individual interests–values–beliefs, on the one hand, and SMO activities–goals–ideology, on the other, are congruent and complementary (Snow, Rochford, et al., 1986). Protesters work to convince others that their definition of what is wrong is reasonable and that their suggested remedies are practical enough to merit listeners' active support (Klandermans, 1988). Local residents are more likely to be available for protest mobilization when they perceive a proposed incinerator in a negative "frame." Alternatively, the more such plants are perceived positively by nearby citizens, the more unlikely it is that opposition groups will successfully recruit widespread support in the area.

Framing has been receiving increased attention in recent years, with special emphasis on the role of central interpretive concepts, or "master frames," in mobilization processes (Snow and Benford, 1992). If framing is something everyone uses to interpret the world by simplifying and condensing it, master frames in collective action do the same thing on a larger scale, with a view toward pulling together activists and social movement organizations by assigning blame for some problem as well as its remedy. Successful framing by challengers is said to depend on diagnostic, prognostic, and motivational aspects. The identification of a problem and attribution of blame for it is called "diagnostic framing"; the proposed solution to the problem is referred to as "prognostic framing"; and "motivational framing" is the "call to arms for engaging in ameliorative or corrective action" (Snow and Benford, 1992). Despite considerable theoretical reflection on frames and framing processes, there have been few empirical applications of the resulting conceptual tools (see Gerhards and Rucht, 1992, for a notable exception). We shall draw upon this literature, evaluating the relationship of differential framing of incinerator issues by individuals and groups in backyard communities, as well as in surrounding counties, to the outcomes of various siting conflicts.

Countermovement Organizations

At some of the sites we examine, supporters of incinerator projects formed citizen advisory committees (CACs). Although stretching the usual meaning of the concept a bit, these were actually countermovement organizations, even if formed in anticipation of the grassroots SMO rather than afterward. Their intended function was to organize enough influential support for incinerator projects to neutralize the expected anti-incinerator mobilization efforts by backyard residents (Konheim, 1986). Industry consultants endorse the early formation of such CACs, and although typically portrayed as neutral fact-finding organizations, they were actually functional equivalents of the countermovement organizations discussed in the equity movement literature (Zald and Useem, 1987).

According to incinerator proponents, citizen advisory committees serve as vehicles for providing citizen input on siting decisions, but opponents typically view such organizations as co-optative mechanisms that help limit the scope of dissent and control local residents. The nuclear industry's earlier relationship to pro-nuclear citizen organizations provided a model for decision-makers in the incinerator industry to encourage the formation of such CACs in communities considering a plant's siting—especially after the mid-1980s, when the anti-incineration movement became more influential. Such technology protests are virtually inevitable after a LULU is announced, focused as they are on opposing the construction or operation of such facilities, and this accounts for proponents' launching a preemptive strike in the form of a CAC even before local residents mobilize the protest itself. Some projects we examine have nominal CACs associated with them.

Specific Variables

We also turned for insights to national opponents and proponents of incineration as we designed our research. Anti-incinerator activist Paul Connett, for example, emphasized the importance of early organized resistance within the target community as a critical variable accounting for many successful citizen challenges. Industry consultant Carolyn Konheim, on the other hand, pointed out various factors that experience had convinced her were critical in facilitating sitings after the early 1980s, when "90 percent of successful plants are there in spite of local citizen opposition." She felt that such local opposition was inevitable and advised proponents to ignore it and "concentrate on winning and retaining a broader constituency outside the backyard area." Konheim also emphasized the importance of "a widespread perceived need for the facility and a local official championing it" and recommended that proponents "establish a citizen advisory committee composed of a broad cross-section of people relying upon a scientific site selection process." She noted that "the sizing of these facilities is very important and larger ones are more difficult to get through," and she warned that "if grassroots citizens groups have been previously active in an area they are likely to be reactivated in opposition to such a project." Konheim also suggested that proponents do all in their power to prevent any referendum on such siting issues.[6] Other informed observers also suggested that the larger the proposed incinerator, the more likely it was to encounter citizen opposition. Some said that size was relative to the volume of the local trash stream and that the critical question was whether the facility would have to accept out-of-county trash in order to operate near peak efficiency. We shall examine these and related variables.

The theoretical relevance of the Cerrell Report—the study by industry consultants that later became an organizing tool for opposition groups—was its listing of variables allegedly associated with communities that are more or less vulnerable to an incinerator project (Cerrell, 1984). Based on previous industry experience as well as a review of social science literature, this report sketched a profile of people likely to be "least resistant" to incinerators and other such projects. As noted, opposition activists used this document as a wake-up call in target areas, claiming that the incinerator industry despised residents as another "Cerrell community."

Grassroots protests against various technologies—modern incinerators, hazardous waste dumps, nuclear power plants, nuclear waste sites, and similar targets—are becoming increasingly common throughout the industrialized world. In these movements, newly perceived risks based upon information and interpretations offered by technical specialists form the basis of an ideological protest frame for environmental activists. Challenging sophisticated technologies outside the ken of many educated laypeople, these protests often depend on maverick experts to interpret the issues and achieve credibility in the eyes of the public. Rather than lining up against targets defending the cultural status quo, as typically happens in equity movements, activists in technology movements challenge an industry attempting to initiate a type of change that they resist or to which they propose an alternative. Citizen activists, however, often confront problems in the legal arena deriving from regulatory agencies' symbiotic links to the industry in question.

Success is typically linked to the elimination of one's antagonist in technology movements. And while there are a variety of debatable criteria for success, few are more credible and handy than whether a facility is sited or not. One of the more interesting characteristics of these movements is that

6. Konheim interview, in the course of which she suggested that these variables were critical in such siting efforts. See also Konheim, 1986.

displacement of one's opponent is typically a sine qua non for success, whereas it is the surest predictor of equity movement failure. Displacement prospects are not nearly as dim for technology movement organizations in such zero-sum struggles.

Taking a cue from the contemporary debate among social movement analysts about the relative importance of indigenous versus outside resources, we first compare residents living in the backyard communities within a few miles of each project in terms of their mobilizing potentials. If indigenous resources were decisive in these technology protests, the backyarders in areas where incinerator projects were defeated should rank higher on such indicators as socioeconomic status, network density, and anti-incinerator framing than backyarders where such plants were built. If the internal resources of these backyard communities are less than decisive in explaining outcomes, then we shall have to rely more heavily upon our fieldwork data in Chapters 4 through 7 in accounting for outcomes.

And how about the broader social settings surrounding the respective backyard communities within which successful challenges occurred? In general, both the existing social movement literature and individuals involved with such siting attempts suggest that the more outside support communities receive in neutralizing proponents the better. In addition, previously successful protest mobilization around any issue in the general vicinity of a siting is viewed as a valuable skill that can be reactivated for an incinerator project, and if the project requires importing wastes into the county, it is all the more likely that it will generate widespread opposition from outside the backyard area. If opponents are able to get the incinerator issue voted upon in a countywide referendum, that is also regarded as boding ill for the project. Incinerators are more likely to be sited, on the other hand, where backyard communities have characteristics that contrast with those just mentioned.

Before turning to detailed analyses of each incinerator project, we focus on the people living within a few miles of the different sites. Chapter 3 compares results from telephone interviews with these backyard residents, addresses some intriguing theoretical issues with relevant survey data, and provides a bird's-eye view of all eight target communities.

3

Eight Backyards: Surprising Survey Results

The initial thrust of contemporary "resource mobilization" perspectives emphasized the critical importance of outside assistance for deprived collectivities with grievances, but subsequent work accentuated the role of the aggrieved collectivity's own internal resources in promoting successful protest. Our design and data address this important issue, focused upon the relative importance of indigenous and outside resources. We defined residents living within three miles of an incinerator site or proposed site as "backyarders," or the indigenous community. Telephone surveys with random samples of these backyarders enabled us to examine the relationship between specific characteristics of indigenous communities and siting outcomes. In subsequent chapters, we expand our focus to include actors and processes outside the immediate neighborhoods of these projects, but in this one we concentrate on survey information from "backyarders" within a few miles of each site.

We expected to find numerous differences between communities where incinerator projects were defeated and communities where, despite organized protest, they were successfully sited. According to the literature already reviewed, people at lower socioeconomic levels are less likely to be successful challengers because they lack the resources available at higher levels. Thus, we hypothesized, backyard residents in areas where incinerators were successfully sited should either have favored the project or rank lower on such socioeconomic status indicators as income and education than residents in areas where such projects were defeated. In addition, the Cerrell Report indicated other demographic characteristics of populations where successful sitings occur—such as being older, being longtime residents, being politically uninvolved, being Catholic, and being Republican.

Contemporary analysts use the concept of "framing" to refer to people's attitudes toward such projects, suggesting that the more positively a particular backyard population framed the issue, the more likely the incinerator would be sited. This framing logic suggests, conversely, that the more negatively incineration was viewed by residents living within a few miles of a proposed project, the less likely any incinerator would be successfully sited in that area.

The critical importance of networks is one of the central emphases in contemporary social movement analysis, and we included a cluster of items in our telephone survey designed to measure various aspects of respondents' involvements with friends, neighbors, anti-incinerator activists, and others. The literature reviewed in the previous chapter predicts that the density of preexisting networks will be directly related to any community's success in mobilizing along the lines of the majority's political preferences. Following such logic, we would expect the network density of any given backyard area to be an important factor in helping explain the success or failure of attempts at organized protest.

The Appendix contains specifics concerning both the methodology and the data. We limit our discussion here to the selection of sites and respondents and to the types of questions used to measure concepts emphasized above. Focusing upon three northeastern states—New York, New Jersey, and Pennsylvania—which were among the nation's leaders in both attempted and successful sitings of such facilities, we considered dozens of relevant controversies in these states before finally deciding upon our eight finalists. In efforts to select theoretically interesting sites, we had extensive discussions both with county and regulatory officials and with national proponents and opponents of incineration. Informed observers warned us there was no such thing as a typical case, or any obvious pattern associated with successful or defeated sitings. We intentionally avoided two extremes that seemed least theoretically interesting: successful incinerator sitings where there was virtually no opposition, and defeats where proponents walked away at the first indication of public opposition. Approximately twenty sites in the three states were closely checked through personal visits and/or telephone calls. Each was in process after the anti-incineration movement matured in the mid-1980s, and all involved various levels of organized conflict. Among our eight cases, six outcomes were virtually decided by late 1989, when we began collecting data—three sitings and three defeats—while the other two (Broome and St. Lawrence Counties in New York State) were still in process. The six cases decided beforehand were:

SUCCESSFUL SITINGS	DEFEATED SITINGS
Chester, Delaware Co. (Pa.)	Woodbine, Cape May Co. (N.J.)
Manchester, York Co. (Pa.)	Dunmore, Lackawanna Co. (Pa.)
Plymouth, Montgomery Co. (Pa.)	Navy Yard, Philadelphia (Pa.)

Although our research design also intentionally included a "likely success" and a "likely defeat," based upon informants' evaluations, we were as surprised as other observers when the Broome County project was defeated by a single vote in the county legislature in 1992. We had already collected our survey data around that site, and we had known from the beginning that such an outcome was possible. Commenting upon the unexpected Broome results, one industry consultant observed: "Maybe that tells us something about momentum in this industry." This instructive reversal meant that we ended up with three successful sitings and five defeats (Figure 1) because the St. Lawrence County project was also thwarted.

A professional firm carried out the telephone interviews with a total of 800 randomly selected respondents from the eight sites. Approximately equal percentages of women and men were interviewed at each, and filter items were used to eliminate both respondents who lived farther than three miles from the proposed site and those who had not lived in the area while the conflict was in process. We interviewed 100 of these randomly selected residents at each of the eight sites.

Ideally, researchers might have interviewed these backyarders about their attitudes toward incineration *before* any such project was announced, and then again at periodic intervals over the course of each particular conflict. In fact, though, such an ideal research design is impossible in the real world. Thus, it is always important to be careful about causality assumptions involving residents' attitudes toward such projects. Some people may modify their attitudes to fit existing situations, even when asked to tell how they felt beforehand, but we have no reason to believe many do so in cases like

Figure 1. Map showing the eight focal counties and incinerator sites in Pennsylvania, New York, and New Jersey.

those in our study. To the contrary, it is more likely that a majority of respondents did not change their survey responses to fit the project outcome in their areas.[1]

The items in this telephone survey asked backyard respondents about a variety of issues—including relevant demographic characteristics, their framing of the conflict as measured by a series of responses to positive and negative statements about incinerators, the nature and extent of their friendship networks, their sources of information during the incinerator dispute, and related topics. The full questionnaire with its response options is included in the Appendix. In the remainder of this chapter, we present and discuss the main results from these telephone surveys, results that seem at first to confirm many of the expectations suggested by the literature reviewed earlier. Upon closer examination, these survey results raise more questions than they answer, preparing the way for the eight case studies that draw upon fieldwork and other relevant data to address unanswered questions.

Aggregate Comparisons Between Sited and Defeated Projects

For these first comparisons (Table 1), we combined responses from the backyards around the three sites where the incinerator projects were successful and contrasted them with the responses from the five sites where similar projects were defeated. We refer to the collectivity of respondents from the five sites where citizen opponents successfully challenged incinerator projects as "Defeated" and to those from the three sites where projects were successfully built despite organized protest as "Sited." In the first section of Table 1 are statistically significant differences between the two groups on all seven measures of attitudes, each in the expected direction (details regarding the statistical methods are in the Appendix). The data reveal that backyarders living in communities where the project was defeated were more opposed to siting. The same backyarders were also *less*

1. A majority did not express either a positive or negative position on the project. A total of 51 percent indicated they had mixed feelings (25 percent), no opinion (16 percent), or were unaware of the incinerator controversy (10 percent). Such findings suggest that large numbers of people did not modify their attitudes to fit these project outcomes.

Table 1 Comparison of attitudes, network, and demographic characteristics of all backyarders

Backyarder characteristics	Sited	Defeated
Attitudes toward incinerator plants		
1. Percent who were opposed to siting of trash plant	18.0	38.0**
2. Percent who believed incinerator would solve trash problem	86.0	70.0**
3. Percent who agreed that trash plant good for local economy	56.0	38.0**
4. Percent who agreed burning better than landfilling wastes	74.0	52.0**
5. Percent who agreed trash plant bad for human health	37.0	60.0**
6. Percent who believed incinerators lower property values	33.0	61.0**
7. Percent who agreed trash plant reduce efforts to recycle	30.0	49.0**
Network characteristics		
8. Percent who discussed incinerator plant with others	59.0	63.0
9. Mean number of different people talked to about the plant	7.9	13.6**
10. Percent who reported their friends had a different position about the plant	35.0	32.0
11. Percent who heard of public meeting about the plant	51.0	66.0**
12. Percent who attended public meeting about the incinerator plant	7.0	15.0**
13. Percent who belonged to any opposition group	3.0	5.0
14. Percent who belonged to groups not involved in incinerator controversy	18.0	14.0
15. Percent who got "quite a bit of information" about the plant from conversations	23.0	33.0*
Background characteristics		
16. Voted for Bush in 1988	49.0	50.0
17. Percent with some college education	41.0	40.0
18. Percent with income greater than $25,000	62.0	56.0
19. Percent who own residence	80.0	78.0
20. Percent who are Catholic	31.0	59.0**
21. Percent who are white	86.0	92.0*
22. Mean length of time lived in area (years)	33.2	33.4
23. Mean age (years)	48.9	48.4

*p<.05
**p<.01

likely to believe the proposed incinerator would solve their trash problem, be good for the local economy, and reduce landfill wastes. They were *more likely* to say that incinerators threaten human health, lower property values, and reduce efforts to recycle. These findings suggest that those organizing protest efforts against defeated incinerator projects were more successful in creating negative "frames" in their own backyard communities than were activists in backyard communities where incinerator projects were successfully sited.

A second master question in the telephone interviews was whether grassroots networks were indeed more dense in backyards where projects were defeated. Contemporary social movement perspectives are somewhat supported by the data in the "Network Characteristics" section of Table 1. There were no differences between the two collectivities in whether respondents discussed the project with others, had friends who felt differently about the project, or were affiliated with opposition or other groups, but there were differences on other network variables. Backyarders in the vicinity of defeated projects talked and networked with more people, were more likely to hear about and attend public meetings about the project, and were more likely to get information about incinerator issues from conversations with other people. Such data offer some support for the notion that backyarders in communities that successfully mobilized against such projects were likely to have more dense and active social networks, but they also suggest that the network differences were formed in response to incinerator challenges rather than from preexisting ties, as assumed by many contemporary social movement analysts.[2]

Were the demographic characteristics of these various backyard communities what the literature suggested on the basis of such outcomes? Although the Cerrell Report indicated that incinerator sitings would meet less resistance in communities with certain background characteristics, the data displayed for items 16–23 in Table 1 hardly support the notion that backyarders in "Defeated" and "Sited" communities differed dramatically in such respects. While the slight racial composition difference is in the direction expected by those who argue that such projects are more likely to be sited in minority communities, the only other significant background difference reverses the Cerrell Report's predictions: defeated projects were more, not less, likely to include Catholics in the surrounding communities. Neither do the data in the table support the common assumption of contemporary social movement theory that collectivities with higher socioeconomic status are likely to mobilize more successfully. We do not find the backyarders who defeated projects to be of higher socioeconomic status than their counterparts in communities where projects were sited.

These aggregate results from the backyarders across eight sites reveal

2. The two collectivities also did not differ with respect to previous protest involvement. These data, together with other comparisons, are in the Appendix.

major differences in their attitudes and framing of incineration issues, some dissimilarity in their networking, but few demographic discrepancies between the defeated and sited communities. We were not surprised to find that backyarders living in the vicinities of defeated projects generally framed incineration issues more negatively and reported more dense network ties. The Cerrell Report's community profile of "least resistant" communities also prepared us to expect the similarities we found in terms of demographic characteristics. In other words, if each of these prospective sites was selected from among others in its respective county because it promised to be "least resistant," they might all be somewhat similar to one another in terms of at least some of the background characteristics. Our next question, however, is whether the differences among "sited" and "defeated" aggregates will hold up upon closer scrutiny of these individual backyard communities. Because such aggregate data may cloak important differences among the three communities where projects were successful, as well as among the five where they were defeated, we next compared the sited and the defeated backyarders with one another, on the same variables as those considered in Table 1.[3]

Three "Sited" Communities Compared

Table 2 includes the same attitudinal variables, but in this case the responses of backyarders in each of the three sited communities are displayed separately. There are statistically significant differences on six of the seven items in the top section of Table 2, and the dominant pattern is that the backyarders in York differ from the other two sites. Those living near the York incinerator were *less likely* to consider incinerators threats to health, *less likely* to believe incinerators would lower property values, *less likely* to say that the building of an incinerator would reduce their community's efforts to recycle, and *more likely* to consider burning better than landfilling. In only

3. We have not included any multivariate statistical models in this book, in an effort to make the survey data as accessible as possible to all readers and to disaggregate these data in a straightforward manner. We have, however, explored several logit models using the variables in Table 1 to determine which are related to plant sitings. Briefly, attitudinal variables (an index created for those numbered 2, 3, 4, 5, and 6 in Table 1), network variables (8 and 11), and background variables (18, 20, and 21) differentiated the two backyard collectivities the same way these variables differentiated them in Table 1.

Table 2 Comparison of attitudes, network, and demographic characteristics of backyarders at sited incinerators

Backyarder Characteristics	Communities: Montgomery Co.	Delaware Co.	York Co.
Attitudes toward incinerator plants			
1. Percent who were opposed to siting of trash plant	26.0	17.0	11.0*
2. Percent who believed incinerator would solve trash problem	94.0	77.0	90.0**
3. Percent who agreed trash plant good for local economy	49.0	57.0	63.0
4. Percent who agreed burning better than landfilling wastes	64.0	74.0	84.0**
5. Percent who agreed trash plant bad for human health	40.0	46.0	22.0*
6. Percent who believed incinerators lower property values	36.0	41.0	22.0*
7. Percent who agreed trash plant reduce efforts to recycle	31.0	43.0	16.0**
Network characteristics			
8. Percent who discussed incinerator plant with others	64.0	54.0	58.0
9. Mean number of different people talked to about the plant	9.7	5.7	8.3
10. Percent who reported their friends had a different position about the plant	42.0	25.0	39.0
11. Percent who heard of public meeting about the plant	60.0	25.0	68.0**
12. Percent who attended public meeting about the incinerator plant	16.0	2.0	4.0**
13. Percent who belonged to any opposition group	3.0	1.0	4.0
14. Percent who belonged to groups not involved in incinerator controversy	22.0	9.0	23.0*
15. Percent who got "quite a bit of information" about the plant from conversations	19.0	24.0	27.0
Background characteristics			
16. Voted for Bush in 1988	53.0	40.0	57.0
17. Percent with some college education	54.0	28.0	40.0**
18. Percent with income greater than $25,000	69.0	55.0	64.0
19. Percent who own residence	78.0	74.0	88.0*
20. Percent who are Catholic	48.0	30.0	16.0**
21. Percent who are white	94.0	65.0	100.0**
22. Mean length of time lived in area (years)	30.2	33.1	36.2
23. Mean age (years)	49.7	47.6	49.4

*p<.05
**p<.01

one case did a county other than York stand out from the other two: Delaware County backyarders were least likely to believe that incineration would solve the county's trash problems. The most remarkable generalization we can make about these attitudinal data is the similarity between Montgomery and Delaware Counties on most variables, and the differences of the York backyarders from these other two. This raises questions about the reasons for these differences in the framing of issues among the three communities where incinerators were sited. We shall address that question with the details from our field data in subsequent chapters.

Network characteristics for the three sites where incinerator projects were built (see the middle section of Table 2) showed no statistically significant differences on five of the eight variables. The differences that did turn up call special attention to Delaware County, where many fewer people had even heard about any public meeting on incinerator issues and where the percentage likely to belong to any type of organization was also lower than in Montgomery and York Counties. These data suggest that the Delaware County opposition either did not hold as many public meetings on the incinerator issue or did not publicize such meetings very well, and also that the backyarders around this project had fewer preexistent network ties. The data also reveal that a higher percentage of Montgomery County backyarders attended a public meeting about incinerator issues.

Our data on demographic characteristics of backyarders in these three communities where incinerators were sited reveal more differences among the three communities than we might have expected. Delaware County best fits the Cerrell Report's predictions regarding target communities' relatively low education, and it also has the highest percentage of minority residents. Montgomery County has the high percentage of Catholics, a characteristic of "least resistant" communities, according to the Cerrell Report. This section of Table 2 also contains some surprising data. Neither the high percentages of Montgomery County residents with some college education nor the percentages of York County backyarders owning their own residences are what we expected, based on the existing literature. The most dramatic differences among the three counties in terms of demographic characteristics are on the percentages of Catholics, ranging from a mere 16 percent in York County to 48 percent in Montgomery County. The most noteworthy generalization about these demographic characteristics is that they reveal more heterogeneity than homogeneity, which suggests that background variables are of relatively minor importance in accounting for these siting outcomes.

What have we learned from this backyard survey data about the three communities where incinerators were still successfully sited, despite opposition? There were as many differences as similarities among these communities, which suggests that the sitings probably took place under different conditions. It is a mistake to imagine that communities where such projects are sited can be glibly profiled in terms of the variables discussed here as if they were monolithic. While the York County backyarders framed the issues closest to the way we might assume for a successful siting, the Delaware County backyarders seemed to have background and network characteristics that better fit what the literature predicted about "least resistant" communities. Montgomery County, on the other hand, resembled a backyard community that might be expected to defeat such a project because of its relatively upscale demographics and high percentage of residents attending a public meeting on incinerator issues—except that it also had such a high percentage of Catholics. We found just as many statistically significant differences (13 of 23) among the three sited communities (Table 2) as we found between our earlier aggregates of sited and defeated communities (Table 1). We next turn to the backyarders in communities that defeated their incinerator projects, to find out whether they were more similar to one another than those we have just examined.

Comparisons Among the Five Defeated Siting Attempts

The five communities where projects were defeated obviously did not have homogeneous attitudes about incinerators or related issues, as a glance at both the size and the number of statistically significant differences among them in Table 3 reveals. In the first item, for example, Broome County backyarders registered surprisingly more opposition to siting a trash plant (64 percent) than either Lackawanna County (24 percent) or St. Lawrence (22 percent) County. We find differences among these backyarders on all seven attitude items in the top panel of Table 3, and a closer examination reveals that the two New York projects in Broome and St. Lawrence differ the most. If these were removed, the remaining three sites would not differ statistically in their perspectives. The overall pattern in the top section of Table 3 is clear: Broome backyarders exhibit the most negative "framing" of incineration issues. They are most likely to oppose a trash plant, to deny that it

Table 3 Comparison of attitudes, network, and demographic characteristics of backyarders at defeated incinerators

Backyarder characteristics	Communities				
	Broome	Cape May	Philadelphia	Lackawanna	St. Lawrence
Attitudes toward incinerator plants					
1. Percent who were opposed to siting of trash plant	64.0	37.0	41.0	24.0	22.0**
2. Percent who believed incinerator would solve trash problem	58.0	67.0	67.0	69.0	90.0**
3. Percent who agreed trash plant good for local economy	22.0	43.0	43.0	40.0	43.0**
4. Percent who agreed burning better than landfilling wastes	40.0	40.0	60.0	61.0	57.0**
5. Percent who agreed trash plant bad for human health	72.0	66.0	60.0	49.0	53.0**
6. Percent who believed incinerators lower property values	78.0	66.0	59.0	52.0	49.0**
7. Percent who agreed trash plant reduce efforts to recycle	68.0	39.0	41.0	45.0	54.0**
Network characteristics					
8. Percent who discussed incinerator plant with others	79.0	70.0	48.0	45.0	72.0**
9. Mean number of different people talked to about the plant	18.3	17.7	10.3	8.3	12.9**

10. Percent who reported their friends had a different position about the plant	24.0	32.0	24.0	48.0	37.0**
11. Percent who heard of public meeting about the plant	79.0	62.0	49.0	58.0	81.0**
12. Percent who attended public meeting about the incinerator plant	19.0	22.0	12.0	7.0	13.0
13. Percent who belonged to any opposition group	6.0	7.0	9.0	5.0	0.0
14. Percent who belonged to groups not involved in incinerator controversy	14.0	23.0	7.0	11.0	14.0*
15. Percent who got "quite a bit of information" about the plant from conversations	38.0	37.0	27.0	26.0	38.0*
Background characteristics					
16. Voted for Bush in 1988	45.0	60.0	54.0	48.0	42.0
17. Percent with some college education	40.0	38.0	37.0	42.0	47.0
18. Percent with income greater than $25,000	59.0	72.0	42.0	57.0	51.0**
19. Percent who own residence	86.0	84.0	72.0	72.0	77.0*
20. Percent who are Catholic	45.0	44.0	71.0	74.0	61.0**
21. Percent who are white	98.0	79.0	83.0	99.0	99.0**
22. Mean length of time lived in area (years)	20.2	25.2	40.9	43.4	37.3**
23. Mean age (years)	44.7	45.3	51.7	54.1	46.8**

*$p<.05$
**$p<.01$

would solve the county's trash problem, to doubt that it might be good for the local economy, and to agree with statements portraying incinerators as threats to property values and disincentives to recycling. At the other extreme, the St. Lawrence backyarders resembled a community where a successful siting should have taken place, displaying as it did a relatively positive attitude toward incinerators and differing from the rest in its belief that incinerators would solve the county's trash problems. Such framing differences between two defeated projects in the same state are puzzling and demand further explanation, which will come with the fieldwork reported in Chapter 7.

The data for the five communities where projects were defeated show more differences in network characteristics than we found in either of the preceding tables. Respondents in Philadelphia and Lackawanna Counties reveal certain similarities in tending toward the negative end of these network indicators: least likely to have discussed the project with others, smaller networks, and least likely to report belonging to groups not associated with the project. Respondents living in Broome, Cape May, and St. Lawrence Counties were more active in discussing the incinerator project with others and were part of larger networks. Backyarders in Broome and St. Lawrence were most likely to have heard about public meetings on their proposed projects.

On background characteristics among backyarders at these five proposed sites, the right-hand column for the third section of Table 3 reveals statistically significant differences on all except the first two indicators (college education and voting for Bush), and a closer look shows that these differences are not attributable primarily to the two New York counties. The data reveal more differences among these five communities at defeated sites than the data in either of our earlier tables: there were only two statistically significant background differences in the aggregate comparisons (see Table 1), and four in those among the three sited projects (see Table 2). Patterns among these defeated communities are not as evident as in Tables 1 and 2. The Philadelphia and Lackawanna respondents are older, more likely to be Catholic, and have lived in their respective communities the longest; the Philadelphia backyarders also report the lowest household incomes. On the other hand, respondents from Broome and Cape May Counties were younger, shorter-term residents and more likely to own their own homes. Cape May respondents also reported the highest household incomes and the most nonwhites. On the basis of these demographic data, the Philadelphia and Lackawanna backyarders appear "least resistant," according to the

Cerrell indicators, while their Broome and Cape May counterparts more resemble the same report's "most resistant" profile. The real story in Table 3, however, is the remarkable variation in attitudes, network variables, and background characteristics among the five communities that defeated incinerator projects.

What generalizations emerge from these backyard survey data about the five communities where incinerator projects were defeated? There were even more statistically significant differences among these backyarders—nineteen of twenty-three, in all—than in the earlier aggregate comparisons or among the three communities where projects were sited. The five backyards where projects were defeated cluster into three general categories on the basis of the survey data examined thus far. The backyarders in Broome and Cape May seem to fit best the profile of communities where attempted project sitings are unlikely to be successful. Their framing of issues, their network characteristics, and their demographics point to active, hostile collectivities relatively united in their opposition. At the other end of the continuum is St. Lawrence County, where the smallest percentage of backyarders opposed the incinerator and where respondents' relatively positive "framing" of the issues is quite surprising in view of the project's defeat. In between these extremes, we find Philadelphia and Lackawanna Counties, with similar framing profiles as well as demographic and network characteristics that are puzzling for communities in which projects were defeated.

The Need for More Information and Explanation

These survey data are both instructive and surprising. Our findings for the aggregated comparisons of the three "sited" communities (Montgomery, Delaware, and York) with the five "defeated" ones (Broome, Cape May, Philadelphia, Lackawanna, and St. Lawrence) generally fit the literature's predictions. But when we decomposed these aggregates and examined the backyard data more carefully, striking differences between and among them emerged.

Without denying the uniqueness of each siting process, we also expected to discover common patterns for the success stories as well as for the defeats. The next three chapters focus upon the details of specific successful and defeated sitings. We save the two New York defeats for Chapter 7. Although the order of the specific individual comparisons is not driven by any compelling logic, there were certain reasons for the sequence of our pairings. In Chapter 4, we focus upon two of the most atypical of these mobilization projects, which also were the projects with the highest percentages of minority members in their backyard samples: Delaware and Cape May Counties. Then, in Chapter 5, we examine two projects in medium-sized Pennsylvania cities: York and Lackawanna. Chapter 6 considers two instructive protest processes in the same metropolitan area: Philadelphia and Montgomery Counties. We held the two New York projects until last, not only because they were both defeated after the others had been decided, but also to wrap the fieldwork analyses up with the details from national anti-incineration activist Paul Connett's own backyard struggle.

4

The Wrong One and the Wild One: Delaware and Cape May

Our first fieldwork comparison focuses upon the processes involved in two distinct incinerator siting attempts: a successful attempt in Delaware County, Pennsylvania, and a defeated project in a rural area of Cape May County, New Jersey. In both cases, the incinerators were targeted for relatively downscale communities. Both counties began serious searches for solutions to their mounting solid waste disposal problems in the early 1980s, as existing landfills were reaching capacity. Incineration was becoming an increasingly acceptable alternative to landfilling, and county authorities experienced few constraints in discussing their waste problems with waste-to-energy vendors, because grassroots opposition to this technology was still in its infancy. By 1986, without encountering more than nominal citizen dissent, both counties had made official decisions to invest in modern incinerators. Within a year, however, opposition emerged in both counties, from quite different sources and with very different results. By the late 1980s, the Delaware County incinerator was under construction, while the Cape May County plant had been abandoned by that area's Municipal Utilities Authority (MUA) officials after an intense struggle with local opponents. Why such divergent outcomes?

Fieldwork Procedures

The fieldwork procedures followed similar patterns at each of the eight sites. Before visiting the area to conduct more detailed interviews, we discussed by telephone what proponents and opponents at each location considered critical issues. Using many of the same questions asked of backyarders (see Chapter 3) as well as open-ended questions about their group's emergence and development, we also collected questionnaire data from activists and conducted focus group discussions. Additionally, a file of local newspaper articles for each site was created. There were thus five related data-gathering procedures at each location: (1) semi-structured telephone interviewing of key proponents and opponents; (2) random telephone surveys of backyard residents; (3) individual and focus group discussions with activists on both sides of the siting issue; (4) questionnaires for individual activists and CAC members (if applicable); and (5) construction of a newspaper and, where available, newsletter file of relevant articles on the controversy. While survey and questionnaire data were collected within a restricted time period at each site, informal interviewing of key informants continued as relevant events occurred. In revising our analyses, we used comments from informants on both sides.

Site Characteristics

A comparison of variables that some informed observers regard as influencing outcomes in incinerator sitings (see Table 4) reveals that the Delaware County plant was more than five times as large as the one envisioned for Cape May and that it was also geared to accepting outside wastes. Cape

Table 4 Characteristics of proposed incinerator sites in Delaware County (Pa.) and Cape May County (N.J.)

	Delaware Co.	Cape May Co.
Size (tons per day)	2,600	500
Out-of-county wastes accepted?	Yes	Undecided
Estimated population within 2-mile radius	30,000	900
Amount of host community annual compensation	$2 million	$2 million
Citizen advisory committee for incinerator?	Yes	No
Any previous organized protest in area?	No	No

May incinerator proponents, on the other hand, left this importation of wastes issue open. Some observers suggest that local citizen opposition is likely to be directly related to the size of any proposed plant, whereas others say that size is less important than whether a plant will receive outside trash in order to keep operating at some critical level. In either case, it would seem that the Delaware County plant would be the least likely to have been built, especially because it was targeted for a much more densely populated area than its smaller counterpart in Cape May. This should have been another negative for the Delaware County plant, because population density is a characteristic said to vary directly with the likelihood of protest. Compensation proposed in return for allowing an incinerator to be built was comparable for the two sites. Table 4 indicates that although a citizen advisory committee did operate in conjunction with the Delaware County siting, that was not the case in Cape May County—and neither community experienced any significant previous protest.

Except for its citizen advisory committee, then, Delaware County's city of Chester seemed a less likely host community than Cape May County's Woodbine because of the larger incinerator proposed for Chester, its acceptance of outside wastes, and its more dense population within a two-mile radius. In fact, however, the proposed trash plant was built in Chester, and a much smaller one was defeated in Cape May. So, we must look beyond the host site characteristics to explain these results.

For that more detailed understanding, we now turn to accounts of the protest processes involved in these divergent outcomes based upon our fieldwork. These data no longer come only from randomly selected residents living within a three-mile radius of the proposed sites; they come from all the major proponents and opponents of the projects, regardless of how far away they lived.

The Delaware County Siting Process

Initial attempts by Delaware County authorities to site a plant in the county's depressed city of Chester encountered opposition, not because of the anti-incineration mobilization that such projects typically precipitated after the mid-1980s in prospective host communities, but because Chester's Mayor Willie Mae Leake and others in her administration wanted to build their own trash plant less than a mile away from the county's proposed location.[1] While the Leake administration was developing the plans for a gigantic incinerator—reported to have been capable of handling 4,000 tons of trash a day—Delaware County officials contracted with the Westinghouse Corporation in November 1986 for a somewhat more modestly sized facility.[2]

County authorities had announced the beginning of their search for a long-term solution to the area's trash disposal needs in 1982, precipitating public hearings on "trash-to-steam" plants by environmental groups in 1984 and prompting Chester's own initial steps toward hosting such a facility through the sale of tax-exempt bonds. It is not surprising that the Leake administration opposed the county's plans, as Leonidas S. Bean, executive director of the Chester Resource Recovery Authority, explained:

> The county project would be competition against the city's own project, and the county's was to be a mass burn facility. Mayor Leake opposed it because it would pour poisonous substances into the air. . . . The city's project, however, involved a refuse derived fuel plant which includes recycling and would be a cleaner burn.[3]

The implication was that the refuse derived fuel (RDF) plant—typically utilizing shredded waste from which heavier, noncombustible items such as glass and metal have been removed—would be safer for residents' health and more environmentally benign than a mass-burn facility designed to burn municipal solid waste with virtually no processing, but there was no scientific support for such an assertion. A more obvious reason for the Leake administration's preference for its own incinerator was that the city of Chester would control all the profits from such a resource. According to one estimate, having its own plant would produce $34 million in profit for the city, but the compensation it would receive from the county facility would be approximately $2 million annually. Commonly viewed as a city

1. The Delaware and Cape May Counties narratives draw upon field notes, interviews with proponents and opponents of the projects, newspaper files, and open-ended questionnaire responses from individuals actively involved in Delaware County's CAC and Cape May County's citizen protest. We are particularly grateful for insights from different perspectives on this Delaware County siting provided by Jim Cronin, Ron Mersky, Leon Bean, Sylvia Diggs, Andi Getek, Kelvyn Anderson, Veronica Barbato, and Andrew Saul. Quotations from individual interviews are identified in separate footnotes.

2. While the size of this county facility was originally estimated to be 1,100 tons a day, both its size and its cost would keep escalating over subsequent months.

3. Interview with Leonidas Bean, Chester, Pennsylvania, January 17, 1990.

in desperate need of industry, Chester was referred to in Pennsylvania Crime Commission documents as "a place where vice activities—gambling, loan-sharking and narcotics trafficking—have served as some of the few stable, revenue-producing, albeit illicit, industries" (Anastasia, 1993).

The Leake administration's resistance to the county's project, however, never amounted to a serious challenge. One of the few examples of opposition after the county chose Westinghouse in November 1986 was a public statement by the Reverend Commodore Harris, a police sergeant in Chester, who claimed that Delaware County authorities acted with contempt for the city in selecting Westinghouse to build the plant without holding any public hearing: "They've been ducking having hearings. They figure they can put anything in Chester because it's poor and destitute, . . . [it's] mostly black, and they figure they're just going to come in here to dump their waste . . . and let the toxic fumes engulf the people" (Hart, 1986:21). Although Harris claimed to be speaking for a social group he referred to as the West End Ministerium, no organized protest against the county plant subsequently emerged from this or any other part of the city. Another individual, Andrew Saul, told of writing "a few long letters to the Pennsylvania Department of Environmental Resources, . . . but that was pretty much on my own. The DER would not consider my challenge because I lived more than five miles from the area."[4]

Asked why there was no significant organized protest against the county plant in Chester, former Westinghouse project manager, Jim Cronin, suggested that residents may have thought that municipal officials would do their protesting for them:

> The city of Chester opposed this plant from the start. . . . And the city's opposition was so public that the people believed the mayor was handling things so they didn't become involved. The city of Chester had even hired a PR firm to oppose the project. It coordinated the appearance at hearings of people with placards, leaflets, and even a casket to protest the siting. . . . The people felt the mayor would take care of everything.[5]

During the same period, the nearby city of Philadelphia was also in the midst of its own incinerator controversy (see Chapter 6). A Philadelphia

4. Telephone interview with Andrew Saul, October 24, 1990.
5. Telephone interview with Jim Cronin, November 1, 1990.

councilman spoke openly in 1986 about the advantages of having his city's refuse handled by the proposed county plant in Chester, and Mayor Leake reacted with hostility in February 1987 to the county's offer to burn Philadelphia's trash in its new plant, "apparently in direct competition with the plant the city of Chester intends to build." Charging that the county's plant was designed to take the money that her administration's incinerator would generate away from Chester, the mayor said that Chester "won't quit until her project is off the ground, offering the city jobs it desperately needs for a stable budget" (Quinn, 1987:18).

During the remainder of 1987 and for most of 1988, the Leake administration was the only publicized challenge to the county's incinerator project. In the spring of 1987, Mayor Leake appointed what proponents of the competing county plant considered a predominantly hostile committee to study that project. In reaction, Steve McKellar, a Chester city council member who happened to favor the county project persuaded Westinghouse, the prospective builder for the county facility, and Delaware County officials to create their own citizen advisory committee. By September 1987 this county project CAC had been established and was headed by Ron Mersky of Chester's Widener University. Mersky emphasized that the county and Westinghouse had already sent applications to the state for the required permits by the time his CAC was established. Commenting on this citizen advisory committee and its relationship to Mayor Leake's more critical committee, he observed:

> Before our committee came into existence, the mayor and her supporters on council had appointed their own committee to look into this plant. . . . They had really been appointed to find that the [county] plant was bad for the city. There was little doubt that this was their main purpose, and there was a feeling that a truly independent committee should be formed. That's why we were appointed. . . . I got the impression that a particular councilman, Steve McKellar, was behind appointing our committee and he wanted us to come back with a favorable report because he preferred the county plant. I'm not sure why he felt that way. . . .
>
> Westinghouse tried to control the information coming to us in the beginning, but we did our own investigation and came back with a fairly strong endorsement of it. . . .
>
> I was the only technical person on the committee. There was also a fairly prominent physician in town, two schoolteachers, a minister,

> and a homemaker—a real cross-section of people from throughout the city.[6]

A year or so earlier, in October 1986, the same councilman Steve McKellar whom Mersky mentioned as being responsible for appointing the CAC for the county plant had turned against the Chester city project after a meeting with a convicted racketeer:

> Christopher Gorbey, special counsel for the [Chester city] trash project, . . . charged McKellar failed to second the motion to vote to hire a project manager for the [city] plant because of pressure from former mayor and convicted racketeer Jack Nacrelli.
>
> McKellar denied the charge. "Jack Nacrelli never told me what to do. He never threatened me. The only person who ever threatened me over this project is Christopher Gorbey," McKellar, accompanied by attorney Clinton Johnson, said. . . .
>
> Gorbey said McKellar had been struggling with a decision to withdraw his support since the councilman first announced Oct. 21—nine days before the resolution was to be approved—that "Nacrelli had just really come down hard on him, and that he threatened to withhold the organization's support regarding his re-election bid next year."
>
> Nacrelli has denied pressuring McKellar. . . . Gorbey questioned why McKellar, who was on a three-man panel which interviewed three qualified prospective firms, suddenly failed to support the resolution after meeting with Nacrelli prior to Wednesday's council meeting.
>
> McKellar left the caucus meeting to meet with Nacrelli at GOP headquarters located across the street from City Hall, causing the regular public meeting to start 40 minutes late, Gorbey said.
>
> "When McKellar returned to the caucus, he told us he had to think of his future in politics, . . . that he had nothing to gain by voting on this appointment and everything to lose."
>
> Both McKellar and Nacrelli acknowledge the fact that they met prior to the council meeting, but McKellar denies the former mayor had any influence over his decision not to support the resolution.

6. Interview with Ron Mersky, Widener Unversity, January 17, 1990.

> "Jack told me he didn't know enough about the project. He said, 'Steve, vote your conscience,' " McKellar said. (Quinn, 1986)

Other members of the county project's citizen advisory committee explained on our open-ended questionnaire items that they were asked by members of the Chester city council to serve and that their work essentially involved, in one respondent's words, "touring the Westinghouse plant in Florida and speaking with officials there, and looking over their permit applications and reports for the county plant." In response to our questionnaire item asking about "very important decisions" made by this CAC, the modification of a transportation route was the only issue mentioned by more than one committee member. "Routing of vehicles was altered after our CAC expressed concern over the proposed route," one wrote.

Within two months, this citizen advisory committee had recommended that the city enter into the host community agreement for the county facility. Mersky told us that the compensation Westinghouse offered was very important: "Money was the main benefit coming to the city. If it were not for the money, our committee would not have thought this was a very good thing for the city at all."

Despite the county project CAC's recommendation, however, the Leake administration refused to sign any agreement at that time. It was later revealed by the Pennsylvania Crime Commission that in January 1988 a former Chester mayor with a criminal conviction, Jack Nacrelli, became involved "to help mediate a host community agreement between the city council and the county" (Hart, 1989).[7] Allegedly, the problem was that bonds had already been sold for the city plant the Leake administration wanted to build.

> Mayor Leake opposed the county project at the time, and was instead pushing for her own plan to build a $335 million trash recycling plant. Westinghouse wanted the city to approve the host community agreement before its plant was built. But city council members were afraid that if they signed it, they might be sued by

7. This Pennsylvania Crime Commission Report was made public only in February 1989 (Hart, 1989). Although Delaware County Council Vice-Chairman Nick Catania had claimed that he "knew nothing about Nacrelli's involvement in the Westinghouse project," the report states that Catania, Nacrelli, and Westinghouse Project Manager James Cronin had indeed gotten together "in several meetings" throughout 1988. Although Nacrelli denied having been paid for his services, he admitted hoping to be paid for representing project contractors in the future.

> people who bought the bonds for the city project. . . . Nacrelli "just articulated to us their state of mind, their feelings," even when city council members weren't present, [Jim] Cronin said. (Hart, 1989)

During the first half of 1988, Leake and her supporters refused to go along with the plans of the county and Westinghouse, despite strong support for the county incinerator from McKellar and some other members of the Chester city council. However, the Pennsylvania Department of Environmental Resources (DER) announced in March that it would grant environmental permits for the county incinerator unless testimony at a public meeting scheduled for April in the 30-day public comment period "uncovered potential hazards that the state hasn't addressed" (LaBarth, 1988a). Before this public meeting, Mayor Leake held an unannounced executive session of the city council in her chambers that some suggested might have been in violation of the law, and Leon Bean, special city representative for the city's trash project, sent a letter to the City Planning Commission questioning Westinghouse's environmental record and the feasibility of having two incinerators in the same city. Westinghouse's Jim Cronin strongly challenged Bean's allegations about his company's record (Taylor, 1988b).

City Councilman Stephen McKellar expressed concern that Mayor Leake's opposition to the Westinghouse plant—in conjunction with serious questions about the way the developers of Chester's own project were operating, and especially about their way of billing the city but sending along no progress reports—might result in abandonment of both projects: "We are fighting Westinghouse for the sake of our own project. If the city continues to fight Westinghouse, sooner or later they're going to have to go away" (Taylor, 1988a).

The major public confrontation between the proponents of the county plant and Chester city opponents occurred on April 27, 1988, at a public hearing sponsored by Pennsylvania's DER. Mayor Leake, Leon Bean, and other opponents of the county plant joined hundreds of demonstrators, some of whom came in costume and with placards, in booing Councilman McKellar and other speakers who supported the county project.

> Hundreds of angry Chester residents crowded into the Chester High School auditorium last night to lodge their protests against Delaware County's intent to build a trash plant in the city. Mayor Willie Mae Leake and state Rep. Robert Wright were among the approximately 700 demonstrators on hand during the emotionally charged public

> hearing that threatened to dissolve into chaos several times. The evening also featured a brief but bitter off-stage confrontation between Leake and City Councilman Stephen McKellar, a supporter of the Delaware County trash project.
>
> Leake and others attended the hearing to demand that the Department of Environmental Resources deny the environmental permits needed for construction of the county's $276 million facility. . . . "The people of Chester have not, and will not, accept the idea of an unsafe and unacceptable mass-burn garbage plant in our city," Leake proclaimed to the cheering audience and DER representatives.
>
> The mayor also charged Delaware County officials with racism, contending that they have conspired to doom the city's own planned resource recovery plant. . . .
>
> Leake, interrupted by applause several times during her speech, admitted that the city's $335 million resource recovery plant is behind schedule, but insisted that it is safer and more efficient than [Delaware County's] facility. . . .
>
> In sharp contrast to her professional demeanor on stage, Leake was absolutely livid later when talking with reporters in a hallway outside the auditorium. "We're a part of Delaware County," she said. "Why do they have to seek other people when we have the funds to build the plant? Why don't they give us the county trash and support a depressed, third-class city like Chester? I can't think but it's racial." (LaBarth, 1988b)

Councilman McKellar was called "Judas" by fellow African Americans as he recommended at the same meeting that the DER approve the county permit, but he insisted to reporters afterward that the whole event was staged by supporters of the city project. "This is not the voice of the community," McKellar insisted. "What you have here is the voice of the vociferous minority. The silent majority is not in attendance" (LaBarth, 1988b).

Within days of this public confrontation, Delaware County pressured the Leake administration by threatening to stop accepting trash from the city of Chester, "a move that could increase the beleaguered city's refuse disposal costs by as much as $1 million a year" (LaBarth and Taylor, 1988). In an editorial the following day, the local newspaper labeled this tactic "economic terrorism" by county officials from the municipality of Media:

> This is a war in which the courthouse gang in Media controls the heavy artillery—access to the county's landfill where Chester has

> been dumping its trash for free. City officials have had to rely on guerrilla tactics such as the hit-and-run raid on last week's DER hearing. . . . Such a move could cost one of the nation's poorest cities up to $1.1 million a year. It would be an atrocity. (*Delaware County Daily Times*, May 6, 1988)

In the midst of these charges and countercharges, Delaware County Council Vice-Chairman Nicholas Catania issued a public prediction that the city of Chester would eventually go along with the county's plans. What Catania did not mention was that ex-con and former Chester mayor Jack Nacrelli was playing "a pivotal role in convincing city officials to approve the county's trash-to-steam plant" (Hart, 1989). The Crime Commission detailed Nacrelli's role in the incinerator negotiations as an example of how he "still controls the city government nine years after he was imprisoned for racketeering" (Hart, 1989). Catania and Westinghouse officials met with Nacrelli several times, according to the commission document, even though Catania denied knowing about Nacrelli's involvement in the project. According to the report (quoted in Hart 1989),

> The significance of this dispute mediation activity of John Nacrelli must be examined in the context of what we know about racketeering in the solid waste industry. In the course of our investigation, we were involved in collecting volumes of information on racketeering in the disposal of solid and hazardous waste. In our sister states—New York and New Jersey—what we found is analogous to the role Nacrelli played in this dispute: racketeers being used to mediate disputes between competing interests. In New York, the Lucchese [La Cosa Nostra] family was used; in New Jersey, the Genovese LCN family was the dispute mediator. In Chester, we find Jack Nacrelli, a convicted racketeer, being called upon to mediate a dispute. The perception that government must rely upon or call upon convicted racketeers to mediate disputes in an industry historically tainted with organized crime involvement serves no other end but to suggest to the citizens of Chester that racketeers still decide public policy issues.

Although at this time strings were being pulled behind the scenes to work out an agreement between Westinghouse and the city, Mayor Leake again refused in early summer to have Chester serve as host for the county plant. It would later be revealed that Catania and Westinghouse officials were also meeting during June with Nacrelli, who acted as the city's representative to discuss the proposed incinerator (Hart, 1989). Despite added incentives offered by Westinghouse in June 1988, including a $1 million advance payment on the host agreement, Leake insisted that the new slant had not convinced her to change her mind: "I cannot sign a host community agreement. We have our own project, and it is moving forward at full speed. I cannot be involved with another project" (Taylor, 1988c). Catania, however, seemed convinced that Mayor Leake would eventually change her mind. The following appeared in a newspaper account in early August:

> Catania said the financial and possible legal difficulties facing Chester's own plant will eventually convince Leake and others to accept the county plant. "The reality is sinking in down there on a daily basis," he said. (LaBarth, 1988c)

County authorities and the Westinghouse Corporation apparently agreed with Catania's assessment of the situation, because they were seeking commitments of municipal wastes from outside municipalities for the proposed incinerator. By August, Delaware County was still trying to make arrangements for approximately half of the 2,688 tons per day of municipal waste the facility would require to operate at full capacity. Early in that month, for example, the county entered into negotiations with Morris County, New Jersey, which was seeking to dispose of between 1,200 and 1,500 tons per day. During the previous month, Philadelphia Electric Company had signed a twenty-five-year agreement with Westinghouse for the utility to purchase all the electricity produced by the plant. Pennsylvania's DER issued waste, air, and water quality permits to Westinghouse in late September to build the proposed incinerator, prompting Councilman McKellar to emphasize the benefits for his city: "The only thing council should do is ratify the host community agreement," he said. "We need that money in next year's budget" (LaBarth, 1988d).[8]

Another meeting among Westinghouse officials, County Council Vice-Chairman Nick Catania, and Jack Nacrelli took place on October 7, 1988 (Hart, 1989). A few weeks later, on October 25, the Chester city council passed a resolution declaring its intent to sign a host community agreement

8. The Westinghouse host community agreement called for the company to pay the city $6 million over the thirty-month construction phase.

with the Westinghouse Electric Corporation. Mayor Leake abstained at this point, insisting on some design changes that would require a new building permit. In mid-November, the city council voted to approve the new design plans, and again Mayor Leake abstained. Ground for the new incinerator was broken on December 15, 1988, and one of those attending the luncheon afterward was Nacrelli. That was the same day the Pennsylvania Crime Commission had tied Nacrelli to video poker gambling rackets during public hearings in Delaware County (Hart, 1989).

Despite a few relatively small citizen protests against the county's incinerator during its months of construction (1990–91), no further significant challenge emerged to threaten the project. On Earth Day 1990, approximately 100 people stood in the rain on the steps of the county courthouse in Media to hear speeches against the trash plant by Andy Saul and other individual opponents. Environmental spokespersons charged that the Westinghouse plant would annually emit several toxic chemicals, as well as high amounts of dioxin, lead, and mercury into the atmosphere (Anderson, 1990), but such sporadic challenges did little to prevent this Westinghouse incinerator from starting to burn trash in 1991.

Mayor Leake's alternative incinerator project never materialized. A few months after we interviewed Leon Bean in early 1990, a Delaware County grand jury accused Bean of "bribery, theft, extortion and racketeering in the shakedown of contractors" working on the Chester city project. Calling this project "a study in administrative bumbling and corruption," the grand jury charged Bean with operating the project as his "private fiefdom" and alleged: "It would be difficult to imagine a public authority which could conduct its business in a less professional or more inept manner than the Chester Resource Recovery Authority under Leon Bean" (Fazlollah, 1991). The grand jury declared the Resource Recovery Authority broke.

Because the issues had been framed by Mayor Leake's administration in terms of jobs and revenue for the ailing Chester economy, it may have been difficult to mobilize city residents against the county plant—based upon the standard anti-incineration arguments. Perhaps this is as much as we can confidently say, because no serious efforts at such grassroots organizing occurred. Nearby communities with significantly higher socioeconomic levels such as Swarthmore, Haverford, Radnor, and Bryn Mawr, however, might have organized in opposition, but they did not. Students at Swarthmore College, for example, had previously caused a multiyear delay on a proposed major highway commonly referred to in the area as "the Blue

Route."[9] The critical importance of the involvement of such activists outside a particular backyard community will become evident in subsequent cases.

How important for this siting was the Citizen Advisory Committee? Westinghouse's Jim Cronin emphasized its centrality and criticized Catania for opposing it:

> The plant would not have been sited without the CAC. . . . Catania, however, was not interested in any type of citizen involvement. The power in Chester is Republican, even though the city is predominantly Black, and Catania felt he could tell people what was going to be done. . . . Having the city of Chester carrying the opponents' ball meant that when the city dropped it, the struggle was over. As soon as we received the permit, Chester officials went along.[10]

A newspaper reporter from the Chester area agreed that local politicians were not responsive: "Politicians in Chester don't answer to the community because people around here are so poor and uneducated. We've five public housing projects, too many people on welfare, and most individuals viewing themselves as at the mercy of the politicians."[11]

The CAC's leader, Ron Mersky, was not as certain of the centrality of the CAC as Cronin. In response to the question "How important was the CAC to the county's incinerator project?" Mersky replied:

> The CAC was not in on this project from the very beginning. . . . Applications had already been sent to the state for permitting by Westinghouse when we got involved. . . . I don't think the CAC was critical because most people in this city were probably very passive about the whole thing. They felt, "If Westinghouse wants to build it, fine!" Those strongly opposed were taking their cues from the Leake administration, so our report would not have any significant effect on them. It received some local press coverage, but not a great deal. . . .
>
> We made our presentation to the Pennsylvania Department of En-

9. Andrea Getek, a reporter for the local newspaper, the *Daily Times,* expressed some surprise that no anti-incineration protest emerged against either the county or city plants (interview, Chester, Pennsylvania, April 26, 1990). Her insights were useful in helping us understand what she called "a unique situation in Chester."

10. Cronin interview.

11. Getek interview.

> vironmental Protection, so it may have had some influence there, but we were probably not vital to the plant's being approved. We had very little, if any, influence on the general public. . . . At least we did an independent evaluation, but I think that if we had come up negative it could have changed things quite a bit.
>
> Although, personally, the way politics works, I think there's a good chance our report would have been buried somewhere if we had handed Steve McKellar a report saying "This plant is horrible."[12]

Mersky laughed as he added these telling comments. He was, as he put it, "the only technical person on the CAC." In their responses to the questionnaire item asking about the CAC's decisions that "turned out to be very important," one said the CAC made no important decisions and the others mentioned the rerouting of truck routes, which Mersky also noted. In response to a related question asking about the most important accomplishments of the CAC, there was no agreement among respondents: one mentioned the establishment of a student scholarship program in Chester, another noted the CAC's "PR effect."

The Cape May County Siting Attempt

Approximately seventy miles southeast of Chester, an incinerator proposed for New Jersey's Cape May County during this same period initially appeared quite likely to be built.[13] It was, as revealed in Table 4, less than one-fifth the size of Delaware County's, targeted for an area with approximately one-thirtieth the population living within a two-mile radius, and offering the same level of compensation to the host community. Although Cape May County is regarded as a relatively well-to-do seashore resort community along the southern New Jersey coast, it also includes some poor rural areas hidden away from its shoreline, and it was in one of these that the proposed incinerator was to be built.

12. Mersky interview.

13. The following individuals, on both sides of the incinerator issue, were interviewed by telephone or in person for insights on the Cape May siting: Valerie Powers of the New Jersey Department of Environmental Protection, Bill Hoffman, John Bielamowicz, Nabil Hanna, Don Hutchinson, Owen Murphy, Ray Rebman, Al Way, and William F. X. Band.

In personal interviews, both proponents and opponents of the county's incinerator plans agreed on the following basic outline of events surrounding this conflict. Recognizing that their state's landfill space was rapidly being used up, New Jersey's lawmakers passed the Solid Waste Management Act (SWMA) in 1975, making counties responsible for their own trash disposal plans by the early 1980s and requiring public hearings in conjunction with such sitings. Cape May County's Municipal Utilities Authority (MUA) began examining waste disposal options in 1982, estimating that its landfill would reach capacity by 1990. After hiring consultants to investigate alternative systems, the MUA decided in 1985 to build a waste-to-energy facility to handle its trash. In September 1986, Cape May MUA officials took a busload of people, including a few reporters, to visit a Baltimore trash plant comparable to the one it was considering for its own county. A few months later, in January 1987, all 29,000 households in Cape May County were sent invitations to a public hearing on a $79 million incinerator that county authorities were planning to commission. At this public meeting, some speakers asked about dioxin emissions from the plant, and MUA officials called such concerns negligible (McCarthy, 1987a). Two months later, in March 1987, citizen opponents held a meeting at which national anti-incinerator activist Paul Connett spoke strongly against incinerators in general and this specific project in particular. His speech prompted one listener to become involved in leading a two-year challenge against the MUA's plans.

The beginning of organized protest against this project came with the public meeting of citizen opponents to incineration on March 6, 1987, four days before all five county freeholders were expected to vote in favor of a plant to burn 500 tons of trash a day in the proposed facility. This plant was targeted for Woodbine, one of the poorer rural areas of Cape May County. Owen Murphy, the local resident who would assume a major leadership role in the successful struggle against the project, explained his own initial involvement:

> It started for me three years ago in March of 1987. I am kind of a hermit, a bachelor then forty-seven years old, who liked to stay home listening to my classical music and reading my books. . . . I do freelance writing, and I used to write about environmental issues—but not exclusively.
>
> In late February, I was over at the post office when the woman behind the counter, a health-food friend of mine, said, "There's a big

meeting next Friday night about the incinerator. Would you like to come?" I said, "I don't even know what you're talking about, Bonnie." She gave me a flyer on it, and for reasons which are still somewhat unclear to me, I went to the meeting, which is really not typical for me because I just don't usually get involved in things like that.

This was Friday, March 6, and Paul Connett was there. I didn't know an incinerator from a microwave. . . . Connett came in and talked about the dangers of mass-burn incineration. . . . He was really inspiring. He could have gotten us to march on Iraq that night. There were about one hundred people in the audience, and he told us, "You've got about four days because on March 10 the freeholders are going to approve this incinerator, and once it's been approved then shortly after that they'll start laying the bricks." . . .

The proposed site, Woodbine, is about ten miles away from here in the poorest section of Cape May County—completely unlike where you are now. . . . It's black and Spanish, populated by fieldworkers who came up this way back in the 1930s and stayed around. It's what passes for the ghetto here in Cape May County. That's always the place where they try to site incinerators because there's little opposition. It's also controlled by one man, Teddy DeSantis, who runs it like a fiefdom. He's the mayor of Woodbine. . . .

Three-quarters of the people in Cape May County have never heard of Woodbine. Poor people on farms struggling to get through the day, who cannot afford to worry about things like the environment, and get controlled politically by one man who rides around with the turkeys and the liquor at Christmas. It's done "ol' style."

So anyway, we had four days to do something, and I had no time to check up on what Paul Connett had taught us about incineration. In that weekend, we were doing things on his say-so, a man who like Shane had ridden into our valley and got it all going.[14]

14. Interview with Owen Murphy, Stone Harbor, New Jersey, August 8, 1990. Murphy explained that he had taught in Belgium, the Virgin Islands, Connecticut, an Atlantic City ghetto school, and elsewhere during the 1960s and 1970s. He then went to Nashville, Tennessee, in the 1970s and ran bars and lounges until an alcoholic crisis in 1977 prompted his return to Cape May County. "After I gave up booze, I found out I was a fairly quiet person. . . . I had all these walls built around me so that people couldn't get in—nondrinker, vegetarian, I like books, and haven't owned a TV set in fifteen years." At the time of the interview, he classified himself as a substitute teacher and a freelance writer. After this interview, he sent the authors a packet of published articles, official letters, and his own notes on the evolution of the Cape May citizen organization. These are drawn upon selectively in the following narrative.

This crucial March 6 meeting with Connett lasted almost four hours. Talking about mass-burn technology, Connett insisted, "They don't know what goes in; they don't know what goes out; and they don't know what to do with the residue" (McCarthy, 1987b). Connett and several other speakers, including Richard Grover, a former county planner in Connett's own St. Lawrence County in New York State, argued for alternative technologies that would involve further recovery of usable materials at the waste plant itself, as well as a large-scale public education program to encourage sophisticated recycling. Connett also showed a video that depicted waste recovery and disposal systems in Europe, "where voluntary 'three can' recycling programs are recovering 'clean organics' as well as recyclable paper, metals and glass" (McCarthy, 1987c). A reporter at this meeting publicly mocked the MUA and county officials, who said they were "impressed" with the plant they saw on their previously mentioned Baltimore trip, and especially with the claims of these officials that they had spoken with neighbors of the plant who had no complaints. "In fact," the reporter insisted, "the MUA tour to Baltimore did not go into the nearby neighborhood of Westport" (Zelnik, 1987a). Other people at the same meeting expressed anger because no freeholders had bothered to attend.

The local citizen protest organization, calling itself the Environmental Response Network, started that evening. Owen Murphy went home and wrote an op-ed piece that he hand-delivered to the *Atlantic City Press* the next day. The article summarized that evening's speeches and urged readers to call each of the five freeholders (names and telephone numbers were listed). Murphy's article also encouraged people to attend the upcoming freeholders' meeting. It concluded: "Your own life and livelihood and those of your loved ones hang in the balance. If you come, wear your warpaint: Tonight, 8 PM, 2nd floor, main library—it's now or never." Although the *Atlantic City Press* refused to publish the article, Murphy paid personally to have hundreds of copies printed, and then he and others distributed them during the day of the meeting. Murphy estimated that he, Ruth Fisher—the veteran environmental activist responsible for getting Connett to come in the first place—and others "worked for about 96 hours" to alert residents.

The freeholders received dozens of phone calls and a number of personal visits from concerned citizens before their March 10 meeting. Freeholder William Sturm, for example, told a reporter that he had received seventy-eight phone calls concerning the incinerator. "He also said he came home to find two persons waiting in his kitchen to talk to him" (Zelnik, 1987a). Two county townships adjacent to the proposed incinerator passed resolutions on March 9—after officials had either attended or heard about the citizen protest meeting—asking the freeholders to "defer action" on the incinerator decision and to investigate other alternatives. A letter from the county health department, combined with such challenges, may have been what finally prompted the freeholders to put off deciding.

> The final shove to a board that doesn't like to be shoved would originate with a senior environmental planner who has been at the county Health Department for less than 21 months. Ironically, Dee Eckert reportedly had been told by the freeholders to attend the Friday night meeting and file a report. She did, and her memo to County Health Officer Louis J. Lamanna repeated two of Connett's key concerns: no trash sorting prior to burn, allowing for the inclusion of hazardous materials, and the creation of hazardous emissions and residual ash.
>
> Lamanna repeated Eckert's points in a letter to the freeholders hand-delivered at about 3:30 Tuesday afternoon. . . . Freeholder-Director Gerald M. Thornton, an incinerator advocate who heads the Health Department, was incredulous: "Health never asked the MUA anything," he told this newspaper. "One person went to Friday's meeting and then went to Mr. Lamanna. He'll hear about this." Describing himself as "somewhat upset," Thornton charged the Health Department never accepted an invitation to meet with the MUA. . . . "The freeholders control policy," added Thornton, "not Mr. Lamanna." (Zelnik, 1987a)

By mid-afternoon on the day of the vote, these combined pressures apparently convinced three of the five freeholders to delay approval of the incinerator plans, and the other two went along to make it unanimous at their meeting later that night. More than 250 angry people showed up at a hall designed to hold 90 at most, prompting a flustered Gerald Thornton to move the incinerator question from the thirty-fourth place on the agenda to first. The freeholders' early vote to delay any incinerator decision for six months somewhat satisfied but also confused the overwhelmingly hostile audience. Ruth Fisher warned those around her in the audience that these officials would simply wait for the opposition to fade away before coming back with a positive decision. It was the largest political crowd Murphy and others present could recall in traditionally conservative Cape May County. Thornton denied Murphy's request to speak at this meeting, so Murphy stood on a table in an outside hallway and gave "the first public speech of my life," in response to Fisher's suggestion that "somebody ought to do something," as people were starting to leave. In the presence of the television cameras and reporters, Murphy warned his listeners that the delay would last only as long as the pressure continued, and he asked people to give their names and telephone numbers for later mobilization purposes. Murphy recalled later:

> That was really the beginning. We got 150 or 200 names. . . . We were launched, and a couple of days later we met at someone's house, about six people (Bonnie Kozak, Ruth Fisher, Linda Pitale, Bob Filipczak, and Owen Murphy). . . . We discussed leadership strategies, and Ruth said she had to care for her ailing mother in upstate New York, so Bonnie and I agreed to assume responsibility. . . . We had decided to call ourselves the Environmental Response Network, and instead of trying to get people to come to meetings we decided to have a phone-based organization. We'd have it so that people can call up and not only talk to us live, but they can also get a recorded message. . . . Somebody else had the idea about a recorded message, but when we checked with the phone company it was staggeringly expensive. So I went over to Radio Shack and found out that I could get a tape machine, and I had the phone company come in and put two new lines in this house in my little upstairs room.
>
> Both these new lines had phone machines hooked up to them. On one, the machine had a three-minute tape, and we had 15,000 cards printed up with that number on it, saying you can call for three minutes of environmental information. Then we had a press release, and we launched ourselves about ten days after that freeholder vote to delay the decision. In the early days, I made new phone messages daily.[15]

Supporters of the project were not pleased with this March 10 decision. In a ten-page letter to Cape May County freeholders, dated March 16, 1987, MUA Chairman William F. X. Band said that while he and others at the authority recognized the "unusual circumstances attendant to your decision, we do not believe this decision in any way represents the best interests of the County." Explaining that "the Authority's technical team" had

15. Owen Murphy said that he spent approximately $3,400 of his own money during the two years or so he was helping lead this anti-incineration fight and that Bonnie Kozak, another leading activist, spent approximately $1,000 (Murphy interview).

decided on "a mass-burn system together with source separation and recycling" for Cape May County, Band's letter insisted that "the selection of mass-burning incineration is consistent with the findings of virtually every other county in New Jersey and the overwhelming majority of local governments nationwide and throughout the world." It went on to "summarize the appalling consequences which will result from your recent decision," including further delay in deciding how to dispose of the county's trash after 1990, higher project costs when and if the freeholders eventually decide to endorse it, and loss of "the county's credibility in the resource recovery industry, among regulatory agencies, and in the financial community."[16]

For the first few months of its existence, the citizen opposition group calling itself the Environmental Response Network (ERN) essentially involved a few active citizens and approximately 200 people on the "membership list" from the public confrontation on March 6. The activists wrote letters to local freeholders, to newspapers throughout the United States in counties with nearby incinerators, to the project editor of the television program *60 Minutes,* and particularly to their local media. The results of these letter-writing activities were mixed: although *60 Minutes* did not respond to his letter for a year, Murphy said he received replies from approximately 100 people in counties from around the United States—for example, Rome, New York; Lewistown, Minnesota; Windham, Connecticut; and Tuscaloosa, Alabama—where his letters to the local papers had been published. He claimed that all but three said "Don't let them build an incinerator in your neighborhood." Initially the ERN did not get much favorable press, although some outside support followed a local newspaper article in mid-April explaining that the new citizen protest organization did not have membership fees, that members used their own money when necessary, and that they still needed such things as a photocopier, blank VHS tapes, video equipment, and even a cartoonist (Lawson, 1987). Murphy said the ERN activists were treated "more or less as freaks for the first few months" by the local media and that a couple of the early ERN activists "dropped out because I was viewed as a loose cannon." Al Way, another early ERN activ-

16. In a brief June 24, 1994, telephone interview, William F. X. Band politely refused to review a draft of this chapter, explaining that he was using his retirement to work on other projects. In the course of the discussion, he surprisingly suggested that the MUA would not have built this incinerator even if Owen Murphy and the ERN had not become involved. This assertion did not seem congruent with the facts of the case, and other supporters of the project, such as Don Hutchinson, insisted that Murphy's involvement was critical for the project's defeat.

ist, explained that some were concerned that Murphy's bombastic style might lead to a lawsuit against their organization.[17]

The focus of the public activity of the Environmental Response Network in April 1987, the first full month of its existence, consisted primarily of attending the monthly meeting of freeholders, where the ERN publicly criticized the freeholders for "not doing your homework," and writing a series of four articles dealing with the mass-burn incineration issue for a local newspaper. In the midst of this series by ERN founders Owen Murphy and Bonnie Kozak an article appeared explaining Murphy's innovative telephone system using "info-in and info-out" telephones—one to enroll new members and pass information to the ERN (info-in), and the other a twenty-four-hour message center that played a three-minute recorded message about environmentally relevant happenings in the county (info-out). Drawing upon a variety of sources, such as "Rachel's Hazardous Waste News" by Peter Montague, Lois Gibbs's Citizens Clearinghouse for Hazardous Wastes (CCHW) newsletters, and the daily newspapers, Murphy created six different messages each week for the first year. Later, he also drew upon Paul and Ellen Connett's "Waste Not," which began as a newsletter shortly after the ERN started.

Unable to get adequate coverage from the local media, Murphy used his "info-out" telephone to promulgate the ERN message. He explained how it worked:

> Every afternoon or evening I'd make a tape. I'd sit down and say "This is the Environmental Response Network, Owen Murphy speaking," and I'd give the date. "We heard today that the incinerator in Canton, New York, had its baghouse filters break down," etcetera, etcetera. And then there might be something about pollution elsewhere, or whatever. I got better at it, and people started to call in regularly. . . . Every time they called it would light up.[18]

Predictably, the Environmental Response Network had early credibility problems, especially with the freeholders, the Municipal Utilities Authority, and the media. Despite a return visit by Paul Connett to speak at its first public meeting on May 10, 1987, and other attempts to attract wider public interest, Murphy noted that the new protest group received little favorable

17. Interview with Al Way, Ocean View, New Jersey, November 14, 1990.
18. Murphy interview.

media coverage for the first few months of its existence. The following excerpt from a May 20 article gives some indication of the cultural climate at the time:

> Mass-burn incinerator foe Bonnie Kozak asked why no freeholder attended the May 10 meeting of the Environmental Response Network, organized to fight the MUA's trash-to-energy plant. "I'm not going to be biased by listening to two extremes," said Freeholder-Director Gerald M. Thornton. "I would not attend a meeting with the MUA to hear its position either. My opinion is that we're looking at two extremes here. . . . " "But this would have been a perfect opportunity for you guys to meet Dr. Paul Connett," said Kozak. Thornton said Connett only had two years experience in the field of solid waste. "Next we'll have an 'expert' from Idaho saying, 'I've been an expert all day,' " said Thornton. "I heard March 10 that RDF [refuse-derived fuel] was the panacea. I don't hear it anymore. Now I hear recycling."
>
> "We've been educated," said Kozak. "I'm not afraid to admit I was wrong." (Zelnik, 1987b)

The attack on ERN activists was coming not only from the county MUA and freeholders, but also from the state of New Jersey's Department of Environmental Protection (DEP). Murphy emphasized the point that "we were fighting the DEP from Day #1 because incineration was the state policy—there was nothing optional about it. The higher-ups were all in bed with the combustion companies."[19] One of the calls to battle by state authorities came in June at the annual conference of the New Jersey Association of Counties at Harrah's Marina casino hotel in Atlantic City:

> The state Department of Environmental Protection yesterday declared war on the "NIMBY syndrome" and self-proclaimed environmental activists who exploit community fears to disrupt solid waste disposal programs. Assistant DEP Commissioner Donald Deieso announced a new push to get correct information before the public with the creation of the office of communications which will be responsible for doing so. The office . . . will also work with individual counties in their solid waste facility siting processes to help counter-

19. From a note Murphy appended to a newspaper article in his files.

> act the NIMBY syndrome, Deieso told county officials. (Jenkins, 1987)

The freeholders added four months to their initial self-imposed July deadline for a consultant's recommendation on how best to deal with the county's trash problem, extending it until after the November 3 election, in which two freeholders were candidates. MUA Chairman William Band estimated that this delay would cost the county between $30 million and $40 million in lost bond revenues. ERN criticisms of the proposed incinerator were addressed by Band in a series of June and July articles—responding to ERN's earlier series. Referring to Donald Deieso's comments, mentioned above, Band wrote:

> Regrettably, the ill-advised, misinformed, and highly emotional information disseminated to the public by the ERN provides a real disservice to county citizens by raising unreasonable fears and dissension that are totally unwarranted. Moreover, the simple-minded and utopian alternatives posed by the ERN only serve to confuse and draw our attention away from the fundamental challenge of selecting and implementing an economically sound, realistic and environmentally safe solution to our growing solid waste management problem. (*Cape May County Herald-Lantern-Dispatch*, July 1, 1987)

Ironically, it was ERN's involvement with mysterious illnesses occurring among people swimming off southern New Jersey beaches and dolphin deaths in August 1987, which, after initially threatening to undermine the citizen group with internal factionalism, eventually gained the ERN the wider base of support they needed in their anti-incineration struggle. Murphy received more than 100 telephone calls from people reporting in excess of 200 cases of illness during August and September. There were also a significant number of dead dolphins turning up along the beaches, and the ERN led the citizen protest against "business as usual" during this period.[20] This broader involvement, however, also caused strife among ERN leaders

20. In retrospect, other activists who eventually became involved with the ERN were quick to acknowledge that Owen Murphy was, for all practical purposes, himself the ERN during these early days. Ray Rebman, an ERN officer who helped run the organization after Murphy, said in an October 1990 interview: "There were times in the early days when Owen was carrying the ball on his own."

and eventually led to a split between Murphy and the others. Three years later, he shared with us his reflections on this process:

> I felt it was very important that we not become just an anti-incineration group. A lot of these groups around the country that have lost have just been anti-incineration groups focusing narrowly on the one issue. Yet they couldn't rouse enough support for it, and I realized early on that if we were ever going to turn this thing down with a vote we needed lots of people. And to get them, we had to be an environmental group involved in a lot of different issues such as the ocean problems, overdevelopment, and neighborhood problems in this area.
>
> Bonnie [Kozak] had two kids and was working two jobs. As a substitute teacher, I could work all day on this—even when I was in school filling in for some teacher by having kids do assigned work. I could do flyers, op-ed pieces, and other things even while substituting, and then I'd come home and find fifteen or so messages which had to be answered on the phone. I'd also have to respond to increasing media requests and get a new phone message on tape for that night.
>
> So when people started to get sick around here that summer—and I also got sick myself, because surfboarding is my recreation—I turned the tapes over to the ocean-dumping issue. We forgot about the incinerator for a while, and that's when the ERN really took off, especially when the dolphins were dying. So we got all kinds of publicity, and Bonnie didn't want to get involved in all that. She wanted nothing to do with anything except low-key anti-incineration stuff. She thought I was getting too wild on the tapes and that I was saying things that were not necessarily provable, so we had a nasty argument and fell out. By August she and John Maslow were gone because they were afraid I was going to get the ERN into a lawsuit. . . .
>
> We had a big ocean rally on Labor Day. I got involved in a lot of these neighborhood fights, and I sort of established my credentials with people all over the county. I'd march with them and demonstrate and give them suggestions about what to do. (I've got a flair for this, a sense of what to do.) So by the end of the summer, the ERN was known and my name was out there. And about three hundred people showed up for our ocean rally here in Stone Harbor, and

> we created this petition for the governor. . . . We coordinated things down this end [of New Jersey] with the Save Our Shores [SOS] originators farther north. . . . We had three thousand signatures by Christmas. . . .
>
> So these involvements really extended ERN credibility far and wide. We were getting letters from children and teachers all over the Northeast saying, "Please send us more petitions." I answered every letter and sent along the requested information—a tremendous spin-off job that I never realized would emerge when I started. Anyway, besides this ocean stuff, we helped stop development of Stone Harbor Manor. . . . We stopped some sandpit mining in Upper Township, helped some people save two hundred houses over on Delaware Bay, and sort of got people in our debt—if you want to put it that way—by doing the right thing, and knowing what to do all over the county. Not with the makers and shakers, but with just ordinary folks. And Bonnie and some of the others didn't want to cast such a wide net, but I knew we had to do it to gain overall credibility. . . .
>
> I don't have a first-rate brain and I don't have first-rate talents, but during this time I had first-rate energy. I worked twelve or fourteen hours a day. I didn't do anything else. . . . As LBJ [President Lyndon B. Johnson] said, . . . "If you do everything right, and if you do the little things, then you'll win." . . . There was a lot of truth in that. If you made sure the press releases went out on time, if you called people and thanked them for writing a good letter-to-the-editor, if you prepared thoughtful tapes, and did all of the other little things which weren't so little, then things would turn out in your favor.[21]

Such attention to detail by Murphy and other ERN activists brought the incinerator issue increasing public attention, forcing both local and state authorities to take clear positions on it. Murphy himself appeared on a popular Philadelphia television talk show on August 26, 1987, mainly to discuss the ocean pollution problem but also to attract more regional attention to the ERN's incinerator debate. By late August, the mayor of Stone Harbor had taken an adversarial position regarding the MUA project, had sent a letter to all municipalities in the county mentioning a recent report saying that people living within fifty miles of an incinerator have an increased risk of cancer, and had asked for more time to review the county's options. On

21. Murphy interview.

the other hand, a state official working with the DEP had sent a letter to nearby Ocean County freeholders insisting that "mass burn facilities are clearly part of the state's solid waste management plan" (McCarthy, 1987c).

Despite the organization's own internal conflicts, ERN activists mobilized increasing local momentum against the MUA's incinerator plans throughout the autumn of 1987. Owen Murphy led a march of approximately 250 people along the beach, protesting continued sludge dumping into the ocean in early September, and Bonnie Kozak presented a petition with 2,000 names asking the freeholders to reject a mass-burn plan they discussed at a six-hour public meeting in the middle of the same month. Al Way, president of the Ocean View Civic Association, was one of the local leaders Murphy worked with to challenge sewage dumping off New Jersey shores. Way, who had also become active with the ERN, told a reporter at the anti–sludge-dumping demonstration: "Local and state officials know the water isn't clean, but won't acknowledge and try to solve the problem because they don't want a bad impression to get out. But . . . my son lives in San Francisco and he's heard about it on the news—it's been on the news all over the country" (McCullough, 1987).

Ignoring the petition and the demands of thirty speakers criticizing the project, the majority of freeholders voted (3–2), on September 17, 1987, in favor of a mass-burn incinerator. At one point in the meeting, which went from 4:30 in the afternoon to near midnight, Freeholder William Sturm expressed frustration with those advocating more serious recycling efforts: "If you believe that the American people are going to recycle 100 percent, you're smoking happy weed—it don't work that way!" In response to a challenger from the audience insisting that building such an incinerator would necessitate the importation of outside trash, Freeholder Gerald Thornton replied that this would never happen: "Let me assure you that there will be no out-of-county waste coming here, not from Philadelphia or from Camden County or anywhere" (McCarthy, 1987d). Widely publicized expert opinions, however, indicated that the county would indeed have to import wastes during the off-season, and Sturm said that such importation "may be wise." The two freeholders voting against the majority, incidentally, were up for reelection the following year.

Instead of giving up in the wake of the freeholders' vote, Murphy immediately demanded a recall election to remove all the freeholders, telling a reporter that 10,244 votes were required to trigger such an election—although he would later learn that a recall election was not possible in Cape

May County. Because the DEP had to evaluate the county's plan and would not be doing so until the following summer, however, Murphy still had time to work against the project. He admitted, though, that he had no hope that the DEP would reverse the freeholders' vote. This was a critical juncture for the ERN; the protest organization might have folded if not for the determination of this single individual who felt quite isolated in his efforts to continue.[22]

As public sentiment against the proposed incinerator increased, local newspapers started to take a more oppositional stance toward it. Editors criticized the freeholders, arguing that the leadership of the county was not responsive to the citizens, and a few local townships passed resolutions opposing mass-burn incineration. Because the freeholders' vote was not the final word on the incinerator, however, activists and editorial writers in the area called for the state's Department of Environmental Protection to defeat the project when it reviewed the county's solid waste master plan. There was a flurry of critical letters to the editor and editorials in local newspapers in the months after the freeholders' vote, and during the final months of 1987 the ERN's support base increased as a result of this growing public awareness on the incinerator issue. One candidate for local office sued the freeholders for voting to permit the incinerator, charging that their vote was illegal because one of them, Ralph Evans, was under indictment by the state grand jury and also because the vote was "arbitrary, capricious, and unreasonable."

In addition to this increasing outside support, activists were also drawn together during this period in working on a common project. They cooperated to design and construct a large and very popular "mass-burn monster" Halloween float, which was entered in a half-dozen parades with leaflets promising that it would be "set ablaze if the mass incinerator is defeated." It brought the Environmental Response Network widespread community recognition and support. And as Murphy and other ERN supporters worked on the float, their cause was not hurt by the publicity the press

22. Murphy subsequently learned that a Cape May County freeholder could not be recalled. Asked whether, in his opinion, someone else would have continued to lead the fight from that point on if he had dropped out after the freeholders voted in favor, Murphy said: "It wouldn't have happened. There would be an incinerator today. . . . I didn't have a copy machine until last year, so I was paying ten cents a page to have things printed—and we were sending things out all over the place. There was also postage, and petitions. . . . After August, it was just me. I was running things from upstairs here after Bonnie quit, until we built the Halloween float when we all came together again and had a good time" (Murphy interview).

gave to the bribery trial of former freeholders and MUA officials during late October. Ignoring this increasing opposition, proponents went on collecting additional information from firms that were interested in building a waste-to-energy facility.

Drawing upon ideas he had picked up networking with activists in other counties and throughout the northeastern United States, Owen Murphy decided to work toward getting the incinerator issue put on the November ballot. Elsewhere citizens were also organizing to resist freeholders and other officials supporting New Jersey's policy favoring the building of incinerators, and Murphy admitted learning from his interaction with them. For example, from an activist in nearby Ocean County, where an attempt had been made to get a referendum on the November 1987 county ballot, he received a response signed by the "President of SAIN" (Stop All Incineration Now) dated March 24, 1988, which read in part:

> The exact language of last November's non-binding referendum in Ocean County on the proposed mass-burn incinerator [was] "Should a mass burn incinerator be built *anywhere* in Ocean County?"
>
> Six municipalities put that question to the voters, after the Ocean County Freeholders refused to do so countywide. . . . Overall, the final tally was 2–1 against mass-burn. . . . Voters in all six towns gave a clear majority of votes against mass burn.
>
> I hope this information will be of help to you. I have also enclosed a copy of a resolution forwarded to the Ocean County Freeholders by Lacey Township, requesting them to authorize a referendum. This, of course, they did not do.

Material like this was used to get a similar referendum on the Cape May County ballot. What the Ocean County freeholders had done as a result of such widespread opposition, however, was to petition the state DEP Commissioner Richard T. Dewling for a three-year extension to allow local officials to assess the performance of other such incinerators, to bolster recycling, and to explore alternative technologies. Dewling and the DEP were not in favor of granting such a delay, and their main concern was "setting a precedent for other counties that are running into opposition to plans for mass burn." One article reporting this challenge stated: "In addition to Ocean County, the most vocal opposition to mass burn in this area has been heard in Cape May County. Environmental groups . . . attacked a decision by freeholders there to construct a mass burn facility at the Upper Township and Woodbine border" (Vis, 1988).

Because Cape May's freeholders had publicly rejected an ERN member's suggestion at their September 1987 meeting that there be a referendum, Murphy and some other ERN activists decided to approach all sixteen townships in the county to try to get a nonbinding referendum on their November 1988 ballots. He reflected two years later:

> We went township to township to try to cajole and convince them to put it on the ballot, low-balling it the whole way, saying, "Look, we just want to find out what the voters think, even though the freeholders and the MUA are going to go through with it anyhow." . . .
>
> We got Dennis Township, which just adopted the resolution from Ocean County, and put their name on the top of it. Once they did that, we took their resolution to all the other townships in the county. . . . We sent letters to seventy-four council members of these different townships. We made dozens of phone calls. We did everything.
>
> Eventually, eleven of the sixteen townships put it on the ballot. The freeholders said, "It doesn't matter. The incinerator is going to be built anyhow, but if Mr. Murphy wants to go through with this sophomoric popularity poll, let him do it."
>
> Well, came the night of the election, and 73 percent of the people here in the eleven townships voted the incinerator down. Every town voted it down—from 2–1 to 5–1, with 23,455 voting no and 9,100 yes. So the next day we're sitting on top of an overwhelming mandate, and Mr. Sturm—the head freeholder and bad guy—says, "Well, this was a nonbinding referendum."[23]

ERN activists threatened the freeholders with a recall if they ignored the referendum results, insisting that the freeholders should respond to the clear message sent by the voters. Mass-burn proponent and county freeholder William Sturm claimed that the ballot question was misinterpreted: "Clearly there is an anti–mass-burn feeling, but in my view the public doesn't understand the question" (Degener and Cannon, 1988). Sturm admitted, however, that the referendum would prompt freeholders to study the issue in more depth.

23. Murphy interview.

By 1989, the former self-described recluse Owen Murphy had decided to run for political office. During the latter months of 1988, ERN leadership was taken over by Bill Noe and Julie Frasca, as Murphy assumed editorship of the group's newsletter and in the spring of 1989 announced his candidacy on an independent ticket for a freeholder position in that November's election. Flyers summarizing his platform told voters it was time "that Cape May had a watchdog freeholder . . . who doesn't take his marching orders from backroom political bosses, . . . knows how to defend our soil, sea, and sky from environmental outlaws, . . . someone with fresh ideas, new priorities." During the same period, the freeholders had replaced two MUA commissioners and voted out the head, William Band. A few days after Murphy declared his candidacy, the Cape May County MUA announced suspension of plans to build the mass burn incinerator in Woodbine, and Band released a five-page press release criticizing the freeholders for making him the scapegoat: "I deeply resent being unjustly pilloried and vilified as a whipping boy for the benefit of political ambitions," he wrote.

Although receiving approximately 8,900 votes, Murphy was unsuccessful in his bid for a freeholder seat in the November 1989 elections. The winning Republicans received 15,000 votes, but Murphy spent only $7,000, compared with an estimated $100,000 spent by the winners, Sturm and Ralph Evans. A new administration had been selected at the state level, however, and this would mean a neutralization of the emphasis on incineration.

By the spring of 1990, New Jersey's newly elected Governor James Florio had put a moratorium on incineration, and Cape May County was seeking some alternative means for disposing of its daily 500 tons of trash. ERN activists declared victory in their incinerator challenge when they heard in the spring that the Cape May MUA had dropped its plans to build a plant, and the same activists began lobbying for the so-called "compost option," which would take out all the recyclables and noncompostable material, shred what was left, and moisten it, "then microbes are used to break it down into soil." The MUA's plans were to have some composting facility operational by the middle of 1994 (Degener, 1990).

Although nothing definite had been decided, plans for burning Cape May's trash at the Pennsylvania Delaware County plant in Chester were being discussed in May 1990. At a public forum held by the Cape May County MUA, two companies proposed composting, and Westinghouse suggested using its own Chester plant to burn it. Many of the 250 people in the audience, however, opposed burning their trash anywhere, and they cheered when ERN member Ruth Fisher asked Westinghouse representative

Bruce King: "Wouldn't it be hypocritical to send [our trash] somewhere else to be burned?" Another environmental activist at the hearing told King that the county would not really be getting rid of its trash by having it burned at the Chester plant: "How can we possibly justify sending our garbage to Pennsylvania only to have air emissions blow back across the river?" Robert Filipczak asked (Lapusheski, 1990a).

After this meeting, Murphy sent letters to Dan Riley, the new MUA chairman who had replaced Band, and also to Dan Beyel, the freeholder who headed the Republican Party in the area. He intended to pressure these decision-makers to eliminate the Westinghouse option by including a packet of materials on alleged Westinghouse scandals, and he threatened to lead a mass mobilization of county residents against any such project. A couple of months later, in July 1990, the MUA commissioners unanimously resolved to enter into negotiations with a New York City composting firm, Daneco Inc., which had facilities overseas in Italy and Lebanon. ERN activists congratulated the MUA on choosing composting over incineration but questioned the authority's choice of a firm (Lapusheski, 1990b). A few months later, the five Republican members of the MUA went to Germany and then to Italy to examine one of the Daneco plants, prompting Democratic candidates for election in November 1990 to charge that the trip was a pleasure jaunt. An October 5, 1990, editorial in the *Atlantic City Press* criticized those officials:

> It is not an imprudent use of funds to send *one* top MUA official to see how those [European] plants are operating. Let's repeat that for county officials with short memories (who seem to have forgotten the fallout over a 1981 junket): *One*:
>
> If a top MUA official can't be trusted to ask the right questions, get the right answers, and evaluate the plants, then he or she shouldn't be a top MUA official. The number of officials going on the trip leaves the strong appearance that they are more interested in seeing the canals of Venice than watching trash rot.

Not only was the incinerator project defeated, but increasing numbers of local residents were now emboldened to look more closely at subsequent technological proposals, as well as at the powerholders making the decisions.

No citizen advisory committee was created by proponents in conjunction with this Cape May siting attempt, but ERN activists were still meeting monthly during the time we were collecting our data, and we received questionnaire responses from twenty-two activist opponents, including Owen Murphy himself, who had by this time moved to Florida with his mother. In response to one of our questions, these activists mentioned a wide variety of "work done for the group" in various parts of the county—all of it related to projects mentioned above: "campaigned for Owen Murphy," "attended freeholder meetings," "participated in rallies," "met with the MUA," "was treasurer for Owen Murphy," "gathered names on petition for nonbinding referendum," "helped build the mass-burn float," "assisted in publishing our newsletter," and so on. Responses to an open-ended question about "turning points in the group's development" almost uniformly mentioned "the countywide referendum victory in November 1988 when voters turned down the plant 3–1," as one put it, but they also emphasized other important events. Additional "turning points" referred to included "Owen Murphy's campaign for freeholder in 1989," "Owen Murphy's televised debate with Freeholder Bill Sturm," and "When Owen Murphy started using the press." One person summarized this "turning point" question quite succinctly: "I felt it was all organized and well directed primarily by one person, Owen Murphy. The organization was solid in following his leadership, but he was the one with the know-how."

Similarly, in responses to a question about "decisions which turned out to be very important," almost all the activists mentioned one or more of the following: the referendum, Murphy's running for freeholder on an Independent ticket, Murphy's debates, and other projects touched upon in our earlier narrative. Two reasonably representative responses to this item were: "When we decided to gather 500,000 signatures from around the nation on a petition to stop ocean dumping" and "Owen's decision to run for Freeholder focused countywide attention on the incineration issue."

Summary and Reflections

Few of the physical and structural characteristics allegedly associated with such sitings were useful in accounting for either the Delaware County or the Cape May outcome. Contrary to predictions, for example, the sited incinerator in Delaware County was considerably larger than its counterpart in Cape May, was situated in a more populated area, and accepted outside wastes. Our field data on the two processes, however, in addition to helping explain their divergent outcomes, also provided contexts within which Chapter 3's telephone survey data from backyarders could be better understood. In view of the extensive grassroots mobilizing effort in Cape May and the virtual absence of such efforts in Delaware County, we are less surprised that the Cape May backyarders were more opposed and more negative in their framing of issues, and that more of them reported having heard about the incinerator and attended public meetings about it.

What theoretical insights can we draw from these first two cases? The Delaware County siting in Chester illustrates a split among political elites, in the absence of effective grassroots mobilization, and emphasizes the importance of issue framing. In addition to the obvious factors militating against successful protest mobilization in this resource-starved area, officials in the city of Chester framed the struggle in terms of local jobs and municipal revenue, instead of using typical anti-incineration arguments against the county's plant. The intention to construct their own incinerator, in fact, meant that anti-incineration arguments were inappropriate in this case, unless raised by activists opposing both projects. The initial framing of issues by local authorities, however, discouraged outside anti-incineration groups from becoming involved. Whether the citizen advisory committee associated with this particular siting was critical is debatable.

If this successful siting in Pennsylvania's Delaware County resulted from a strange convergence of factors, the defeat of the second incinerator in New Jersey's Cape May County was also atypical in some respects. An energetic entrepreneur living outside the "backyard" community used an innovative telephone technology to frame issues, and his unique blending of networking with issue-framing was ultimately responsible for this defeat. Unsuccessful at first, he created networks around other issues, such as ocean pollution and neighborhood problems, and subsequently mobilized those networks both during the anti-incinerator referendum campaign and during his own run for political office as an anti-incineration candidate.

Based only upon these two cases, we might conclude that the successful siting in Delaware County depended on the absence of an independent grassroots protest organization framing issues in an anti-incineration context, especially in the face of a relatively unified county power structure and vulnerable local officials. Alternatively, on the basis of the Cape May data, we might expect opponents to defeat such projects with a broad-based grassroots protest organization that includes a variety of issues in its frame and hangs around long enough for a split in political elites to occur. But how do such generalizations stand up as we consider additional cases?

5

A Convenient Siting and a Quick Defeat: York and Lackawanna

Two medium-sized Pennsylvania counties that experienced anti-incinerator protests are the focus of our second comparison. The outcomes were a successful siting in York County and a thwarted attempt in Lackawanna County, and our central question is why one plant was built and the other was defeated. Both struggles took place during the latter part of the 1980s. As is typically the case, county officials initiated the York search for a site on which to have a trash plant built, announced the location in 1985, and completed construction by 1989. In Lackawanna County, an individual businessman—not the county—championed the project, which was first announced in 1988 and then canceled within less than a year in the face of organized protest. As in the preceding chapter, we compare certain physical and structural characteristics of these two proposed projects and then analyze the dynamics of the processes.

Site Characteristics

The similarities between the York and Lackawanna site characteristics (Table 5) are remarkable. Both projects planned to accept outside wastes,

Table 5 Characteristics of proposed incinerator sites in Lackawanna and York Counties (Pa.)

	York Co.	Lackawanna Co.
Size (tons per day)	1,344	1,000
Out-of-county wastes accepted?	Yes	Yes
Estimated population within 2-mile radius	30,000	25,000
Amount of host community annual compensation	$300,000	$2 million
Citizen advisory committee for incinerator?	No	No
Any previous organized protest in area?	No	No

and the host communities were promised attractive compensation packages. Neither protest had a citizen advisory committee associated with it, nor had there been any previous organized protest by local residents in either area. The proposed York and Lackawanna plants were each of moderate dimensions and targeted for medium-sized urban areas. The higher compensation package in Lackawanna County should not be overemphasized, because it was never finalized; some critics insisted that it was greatly inflated from the start. These initial data suggest two quite similar projects, providing no obvious clues to why the York project was a success while the Lackawanna project was defeated. For that, we turn to our fieldwork data.

The York County Siting Process

York County's incinerator was sited only after a protracted procedure that, according to one centrally involved official, was overtly driven by politics.[1] In a surprisingly candid telephone interview, this county's assistant director of planning, Joseph Hoheneder, discussed the details of this siting on Black Bridge Road, after eighteen months of debate, in an industrial area at the eastern edge of Manchester Township:

> There were several possible sites for the York County incinerator, but poor coverage by local TV eliminated each of the original ones.

1. The following individuals on different sides of the controversy provided insights that were helpful in reconstructing the process. William Ehrman of the York County Solid Waste Authority, Randy Curry, Manchester Township Manager David Raver, York Assistant Director of County Planning Joseph Hoheneder, Attorney Mark Lohbauer, and especially citizen activist Beatrice Weitkamp. Quotations from individual interviews are identified in separate footnotes.

> . . . The Hanover site would have been better scientifically, for example, and Red Lion Borough actually wanted the incinerator but had some technical problems. . . . There were also political difficulties at each of the other early sites, and only because of that was the Black Bridge Road site snuck in at the last minute. . . . Black Bridge wasn't as openly discussed as the others, and in my opinion it was the worst site but became the "best" only because it was the most remote and offered the least political resistance.[2]

Hoheneder explained that the extremely large York County Solid Waste Advisory Committee (SWAC) did not even have this Black Bridge site on its original list of potential sites. Acknowledging that York's SWAC was not a genuine citizen advisory committee, he viewed it as merely "legitimizing a very difficult decision." The Manchester Township Board of Supervisors accepted the SWAC decision in October 1985.

While anti-incineration activists commonly allege that political power figures much more prominently than scientific information in such sitings, it is remarkable to find an official who is involved in such a process candidly admitting the priority of political over scientific considerations. The York County Solid Waste and Refuse Authority had begun working on alternatives to landfills in the early 1980s, and, as its final report notes, landfilling—the common method then used to handle solid waste in York County and elsewhere—was "experiencing problems due to a lack of space and environmental concerns, particularly ground water pollution" (York County Planning Commission, 1985:xiv). A Solid Waste Advisory Committee that included a wide variety of representatives from various municipalities, the planning commission, environmental groups, business and industry, landfill operators, private waste collectors, chambers of commerce, and the state DER was formed. Hoheneder, himself a member of this more than 100-person "committee," emphasized SWAC's importance in the whole decision-making process. Manchester Township Manager David Raver referred to it as "a lightning rod for the project from the start."[3]

Even before a Pennsylvania law, Act 101, enacted in the latter 1980s gave counties the jurisdiction previously belonging to townships and boroughs in deciding on the disposal of their own solid waste, York County Board of Commissioners had authorized the York County Planning Commission to

2. Telephone interview, with Joseph Hoheneder, October 12, 1990.
3. Telephone interview with David Raver, August 24, 1990.

enter into agreements with municipalities granting the county the authority to prepare a revised solid waste disposal plan. Fifty-eight of the seventy-two municipalities in the county agreed to participate. Although officials in a few communities expressed some interest in accepting the incinerator in return for significant compensation, most resisted. The following excerpt from a newspaper editorial in the *Harrisburg Patriot*, April 16, 1985, was written before the Black Bridge site was even being considered and when Springettsbury seemed likely to be the host township selected by the York County SWAC:

> Officials in Springettsbury Twp. do not want to see their municipality become the site of York County's proposed incinerator. But they are prepared to accept it for a price.
>
> Centrally located and on the outskirts of York, Springettsbury Twp. contains two of five sites under consideration by the York County Solid Waste and Refuse Authority. . . . The negative response of officials in Springettsbury Twp. is the normal reaction one would expect of any community presented with the prospect of serving as host for society's refuse. But at the same time the officials demonstrated an uncommon degree of political maturity, practicality and community self-interest. While indicating they would rather see it go somewhere else, township officials made it clear that if the incinerator was forced on the community, they expected the community to be paid for its troubles. Supervisor Robert A. Minnich suggested a $2-a-ton surcharge would be about right and "would go a long way toward reducing or eliminating the real estate tax for municipal purposes."
>
> Simply put, township officials propose to make the best of a difficult situation by turning a problem into an asset. And in the process they are providing a realistic model of how other communities and their officials can best respond to projects that are undesirable but also unavoidable.

Quite different from a citizen advisory committee such as Chester's six-person panel, the massive SWAC was the primary siting mechanism in this case. It was a major component of the organizational process county officials used, because it included "elected and appointed representatives from every municipal jurisdiction in the county," as well as a wide range of local environmental groups (York County Planning Commission, 1985:2–1). The minutes of the meetings report two key decisions the committee made from the autumn of 1984 to October 1985: (1) the selection of alternatives for solid waste disposal was changed from landfilling to the building of an incinerator (referred to in the minutes as "the resource recovery option") at the November 29, 1984, meeting, and (2) the surprising addition and endorsement of the Black Bridge site for the incinerator at the October 24, 1985, meeting, even though it had not been mentioned in previous meetings. In their letter indicating that they had met on October 22, 1985, and had "voted to accept the Authority's proposal," the Manchester Township Board of Supervisors referred to a September 16, 1985, letter they had received from William Ehrman of the Solid Waste Authority "concerning the York County Solid Waste and Refuse Authority proposal to locate a Refuse to Energy Facility along Black Bridge Road in Manchester Township" (York County Planning Commission, 1985:D-1). The minutes for an October 24, 1985, meeting include the following account of a challenge to the Black Bridge site:

> Mrs. Weitkamp asked why the Black Bridge site was not brought up at previous SWAC meetings. Mr. Lerew stated that this particular site was first considered after the last SWAC meeting. The site was not announced publicly until the Solid Waste and Refuse Authority had a chance to complete their study of the Black Bridge site. Mrs. Weitkamp submitted petitions from the residents of the Pleasureville area indicating their opposition to the Black Bridge site. Mrs. Spillman will be submitting petitions from residents of Woodland View Drive. . . .
>
> There is one essential characteristic that the Black Bridge site had and that is the elected officials of Manchester Township have accepted the site, unlike the municipalities where the other two proposed sites are located. Mr. Crowl made a motion that the Solid Waste Advisory Committee approve the Black Bridge site and make it part of the Plan. . . . Mr. Goodlander seconded the motion. The motion was unanimously approved. . . .
>
> Mrs. Glass then commended Manchester Township for taking the step to accept the Black Bridge site. (York County Planning Commission, 1985:C-74 to C-76)

Other potential sites were thus eliminated, and after its October 1985 meeting the SWAC supported the Black Bridge site, which was located at the eastern edge of Manchester Township. The fact that this area was zoned "industrial," and especially that the prevailing westerly winds promised to carry any smoke, odors, or airborne debris from the facility away from Manchester communities, neutralized most objections from the township's own residents. But in the adjacent township to the east, Springettsbury, some residents did object to being downwind from such a plant.

The one SWAC member who objected to but did not vote against the Black Bridge site at the committee's October 1985 meeting, Beatrice Weitkamp, became the leader of the opposition group from Springettsbury that subsequently called itself ROBBI (Residents Opposed to the Black Bridge Incinerator). She had been on the committee since 1984 as a representative of another environmental group, SAVE (Springettsbury Area is Vital to the Environment). A middle-aged woman with mainstream values, Weitkamp expressed no radical ideas about challenging the siting herself, nor did she encourage such notions in others. One of ROBBI's first public meetings, for example, occurred in the spring of 1986 and involved a joint forum with a county landfill owner who also opposed the incinerator because it would take away his business. This combination of an emergent environmental group working with a landfill operator against an incinerator raised eyebrows and alienated some potential ROBBI supporters:

> The 30 members of ROBBI, who all live in Manchester and Springettsbury townships, are "very uptight" about the planned $130 million incinerator, according to group president Beatrice Weitkamp. "We'll be breathing the smoke off that 200-foot stack," Mrs. Weitkamp said. "And the fly-ash is very toxic."
>
> However, not all residents at the forum spoke in opposition to the incinerator. Two members of the area SCREAM group—Save and Conserve the Resources Around Modern (Landfill)—objected to much of the information presented.
>
> "It's a lot of lies," said Elaine Kaufman, York RD 9, whose property is adjacent to Waste Management's Modern Landfill. (Eib, 1986)

Critics argued that ROBBI neutralized potentially sympathetic residents at the beginning of its struggle by this public appearance alongside a landfill operator whom environmentally sensitive fellow citizens regarded with suspicion or worse.

ROBBI's efforts, however, did receive support from the Pennsylvania

Chapter of the Sierra Club which released the following statement on May 31, 1986:

> The executive committee of the Sierra Club, after a tour of the proposed Black Bridge Road site for a York County solid waste incinerator, opposes construction of the incinerator at that site because of its proximity to the surrounding residential development and because of the scenic nature of the area.

Neutralizing such endorsements were newspaper editorials like the following, which promoted incineration and criticized ROBBI, without explicitly naming that "citizen group":

> The York County Planning Commission will tell the state Department of Environmental Resources that it endorses plans for a waste-to-energy incinerator on Black Bridge Road in Manchester Township. Good for the commission.
>
> But the commission has some reservations about the plans. That's OK, too.
>
> The commission and York County Solid Waste and Refuse Authority are both serving the people, but from different perspectives. Ironing out differences between them may be the best way of arriving at what's proper. . . .
>
> The planning commission is right to be concerned about the issues it has raised. But none of them are insurmountable. None of the issues should delay construction of the incinerator by a whole year as a citizens' group has demanded. (Editorial in the *York Daily Record*, June 17, 1986)

ROBBI sought only a year's moratorium on construction of the incinerator, rather than its abandonment. Specifically, the group expressed concern about the incinerator's size, its potential for "cancer-causing emissions," and the state of mass-burn technology, but it did not really seek to eliminate this project from the beginning. The Sierra Club's position was obviously less compromising.

An indication of the difference between York County's ROBBI and the more aggressive Cape May County Environmental Response Network (see Chapter 4) is symbolized by the short announcement in a York newspaper about a visit to the area by Paul Connett, the anti-incineration activist who had sparked Owen Murphy's involvement in challenging the Cape May project. Under a small headline reading "ROBBI Will Travel to Elizabethtown to Hear Speaker" appeared the following in the *York Dispatch* on September 10, 1986:

> A local group will attend a lecture and workshop on the hazards of trash incineration this weekend. Members of Residents Opposed to the Black Bridge Incinerator will attend a lecture by Dr. Paul Connett, an internationally renowned incineration lecturer, at 7:30 p.m. Friday in Elizabethtown College's Espenshade Hall.
>
> The Lancaster County Network of Citizen Groups is sponsoring Dr. Connett, whose workshop will be from 10 a.m. to noon Saturday. For further information, call. . . .

In other words, rather than inviting Connett to come and speak against the York incinerator, ROBBI notified the local paper that some of its members would attend one of his appearances elsewhere—sponsored by a group from an adjacent county. This is a remarkably low-key approach to grassroots protest mobilization. The news release issued by this neighboring protest group for Connett's talk and workshop on September 12 and 13 read, in part:

> In response to the need to provide qualified expertise concerning the issue that has caused much citizen concern, the Lancaster County Network of Citizen Groups is sponsoring Dr. Connett's Friday night lecture and Saturday morning workshop. . . . Dr. Connett will be speaking on the subject of "Mass Burn Incineration and Alternatives to Solid Waste Management." . . .
>
> Dr. Connett has done considerable research into the toxic effects of mass burn incineration. . . . He has been involved with the incineration issue throughout the United States and Europe. Upon leaving Lancaster County, he will be traveling to Japan for the 6th International Symposium on Dioxin where he will be presenting his research findings entitled "Dioxin Uptake in the Food Chain."

At their next monthly meeting, one of the same outside group's members spoke to ROBBI supporters about cancer risks from dioxin contaminants from incinerators:

> Communities near incinerators that burn municipal refuse can become cancer-risk areas, a Lancaster county environmentalist warned last night. Dr. Thomas Grosh, a Marietta dentist, told about 40 people at the Ramada Inn here that dioxin contaminants emitted from incinerator stacks can cause cancer. He said the incinerator to be built here could be a health threat to residents.
>
> Grosh, a member of the Lancaster County Network Opposing Incinerators, was a guest of Residents Opposed to the Proposed Black Bridge Incinerator. . . .
>
> Construction of the $130 million incinerator is expected to get under way late this year. (McLaughlin, 1986)

In addition to regular monthly meetings with guest speakers over the next year or so, ROBBI concentrated on litigation in its opposition efforts. To raise money for its legal fees and telephone bills, this York grassroots protest organization sponsored a balloon launch, auctions, tree sales, and an anti-litter campaign. The launching of 200 balloons took place on November 9, 1986, and was changed from the initial Black Bridge incinerator site to the more accessible parking lot of the nearby Howard Johnson's Motor Lodge because of inclement weather. ROBBI also joined with other critics in raising questions about an eight-day excursion to Japan during the fall of 1986 by York Solid Waste officials to inspect several trash-to-steam plants while staying in one of Tokyo's most expensive hotels.[4]

At the DER permit hearing on the incinerator in January 1987, ROBBI called upon Paul Connett to challenge the project.

> "This is the high-tech booby prize for the throw-away society," said Connett. "You should make sure you've exhausted your options and you should put off this decision to build a white elephant."
>
> Connett said dioxin emissions from the plant would increase the risk of cancer to area residents. Additionally, dioxin would concentrate in the food chain, increasing the levels of the lethal chemical, he said.
>
> Steven Schwartz, a consulting engineer to the authority, disputed that. Schwartz has studied the possible health risks from the plant, a study not required by the state, but ordered by the Solid Waste Au-

4. Taking local decision-making officials to some attractive part of the United States or to another country on an official tour of allegedly relevant incineration facilities is common enough practice among vendors that anti-incineration critics consider it a bribe.

> thority. "This plant is nowhere near violating [federal] standards. It isn't close to being a problem," he said. (Argento, 1987)

Subsequent monthly meetings of ROBBI during the spring of 1987 focused on serious recycling as an alternative to incineration, with the showing of videotapes created by Paul Connett and his co-workers on German efforts in this regard.

The main problems for the forces supporting the incinerator during this period derived more from bureaucratic red tape than from ROBBI's efforts. Contract specifics, the possibility of an alternative incinerator being built, and bond ratings were critical problems for officials supporting the project during the spring of 1987.

> ROBBI will oppose the solid waste and air quality permits DER is expected to issue within 30 days. Beatrice Weitkamp, president of the loosely-organized ROBBI group, said her group is worried about the health impact from the ash created when trash is burned.
>
> "It will fall flat on its face," she said of the project. "It will be a heavy load on tax money."
>
> Dennis Marcott, project manager for Westinghouse, said DER has strict standards for pollution. "We will be under the limits set by DER," he said. "In some cases, significantly under." Marcott said no visible exhaust can be seen coming from the incinerator. (Smith, 1987)

ROBBI's monthly meeting in April 1987 had Walter Haang, a chemist and director of the New York Public Interest Research Group (NYPIRG), as its guest speaker. Before that meeting, ROBBI members put flyers on windshields in parking lots and along the street, reading in part:

> ATTENTION YORK COUNTY RESIDENTS . . . DID YOU KNOW?
>
> That your garbage collection bill will more than triple if an incinerator is built? That we will still need landfills for the highly toxic ash it would produce? That furans and dioxins will be polluting our air, soil, and water from the incinerator stack? That plastic milk jugs, soda bottles, etc., can be recycled into many useful items? That all dry paper, cereal boxes, junk mail, magazines and papers can be

> made into paper towels, tissue products or fuel for wood stoves? That we could build about 15 recycling plants for the cost of one incinerator. . . .
>
> COME OUT AND LEARN THE TRUTH ABOUT INCINERATION from a scientific expert who investigates public health and the environment, who is the Director of NYPIRG in New York. This author of "The Burning Question" will speak and show slides . . . THERE'S NO CHARGE!

According to Beatrice Weitkamp, most such ROBBI meetings were poorly attended because opponents "had no chance from the beginning, but if Springettsbury Township had helped earlier the siting might have been stopped."[5] Manchester Township Manager David Raver said he agreed with Weitkamp that "Springettsbury's support would certainly have given added credibility to ROBBI's protest at this time."[6]

The DER announced on May 13, 1987, that permits had been issued to the York County Solid Waste and Refuse Authority for construction of the $130 million incinerator. The newspaper story announcing this development also included the following ROBBI threat of legal challenges:

> Although construction of the facility appears a certainty, Beatrice Weitkamp, president of Residents Opposed to the Black Bridge Incinerator, said her group has "not given up the fight."
>
> "We won't take this lying down," she said. "I can't give any specific strategies right now, but I've been in contact with our attorney." Mrs. Weitkamp said DER is ignoring the facts.
>
> "The guidelines in DER are not stringent enough," she said. "At this site there would be a subversion of air, which would, in turn, trap the toxins. I don't care how state-of-the-art this incinerator is." (Frassinelli, 1987a)

Alternatives Weitkamp mentioned in this interview included sewage separation and composting, involving layering wet trash and dirt under the ground.

5. Interview with Beatrice Weitkamp, Springettsbury, Pennsylvania, November 15, 1990.

6. Telephone interview with David Raver, June 23, 1994, after Raver had read an early draft of this chapter.

Despite ROBBI's appeal questioning whether dioxins and furans from the plant would be tested frequently enough, and insisting that ash disposal was not adequately addressed, construction of the incinerator was actually pushed ahead two weeks, "in a move Westinghouse Electric Corporation officials hope will ensure completion of the project within two years" (Compart, 1987). In early August, Smith-Barney Inc. committed to purchase and then resell debt service guarantee bonds totaling $130 million, prompting Lloyd Lerew, chairman of the Solid Waste Authority, to comment: "As far as the Solid Waste Authority is concerned, the bonds are sold and the interest rates are known" (Frassinelli, 1987b).

Construction on the York incinerator began in October 1987, and Westinghouse Electric Company said it expected to complete the project late in 1989 or early 1990. An instructive summary of the York siting process from the vendor's perspective is provided by the following excerpt from a trade journal article entitled "Resource Recovery," written by a Westinghouse spokesperson and published in the late 1980s:

> Survival [of "the resource recovery industry"] in the 1990s will require playing by a new set of rules, with new realism. . . .
>
> That great American philosopher Yogi Berra summed up the new realism perfectly when he said "the future ain't what it used to be." So we must look at the future . . . in a new way and with a broader perspective. Solid waste officials, consultants, and waste-to-energy vendors who succeed in the 90s will be those who can perform within this broader perspective. . . .
>
> The 90s will demand a broader view—including the ability of potential vendors to meet existing and emerging recycling demands and environmental criteria. . . .
>
> In York County, PA, for example, Westinghouse is nearing completion of a 1,344 tpd resource recovery plant. We received a full-service contract from the York County Solid Waste and Refuse Authority in November 1985. The Authority realized from the very beginning that the plant would not be a success unless it was part of an integrated solution to the region's waste disposal needs. They planned ahead for recycling and an environmentally acceptable landfill.
>
> The Authority also knew that the plant would not be a success unless it too was environmentally acceptable to its neighbors. So the Authority asked us to join with them in meeting with the local Sierra

> Club to discuss our plans and exchange views. Although some of our engineers were skittish about the idea, we had a productive meeting. The Sierra Club took no action to thwart our plans, and this helped the plant get approved by the Pennsylvania Department of Environmental Resources. The meeting also soothed much of the early local opposition to the plant before it could fester and grow.
>
> Cooperate (even beyond strict legal requirements). . . . And search for high quality solutions. . . . Take the time to build coalitions with responsible public interest groups. . . . (Pollier, 1989:5–7)

An alternate interpretation of these processes would emphasize that ROBBI never became an organizational carrier of the local Sierra Club's early opposition to the siting itself, as noted above. Our second case in this chapter also involves an early Sierra Club challenge, revealing what happens when a local protest organization becomes such a carrier. ROBBI switched to a Philadelphia attorney, Mark Lohbauer, because according to Weitkamp the group was not satisfied with the advice and some of the legal arguments their first lawyer had developed, especially after they proved ineffective in delaying construction. The announcement in the local newspaper informing the public of this switch also included the following: "ROBBI is soliciting funds from the public for its fight," with donations to be sent to Weitkamp's home address. Predictably, few readers expressed any interest in the organization or, especially, in helping support it.

In January 1988, ROBBI's new attorney appealed the state permits issued for the incinerator, claiming he was confident that the state Environmental Hearing Board would "set aside the permits for the construction and operation of the incinerator." According to the *Harrisburg Sunday Patriot-News*:

> "There is no reason at all for this plant to be built," Lohbauer said. "It creates much bigger problems than it will ever solve for the environment and the health of the people of York County. I am confident the board will see that and revoke the permits."
>
> The setting of a hearing date is at the discretion of the Environmental Hearing Board, Lohbauer said.
>
> Meanwhile, construction of the incinerator is continuing, says Dennis R. Marcott, Westinghouse project manager. . . . All excavation work has been completed, sewage and drainage pipes have been installed and 4,000 cubic yards of concrete have been poured to form part of the plant's foundation. . . .

> "One major attack on the permits will be based on the health risk assessment study done for the project," Lohbauer said. . . . "The DER permit was based on the Malcolm Pirney study, so it is based on false information." (Gleason, 1988)

ROBBI attorney Lohbauer insisted that, if the group won its appeal, incinerator construction or operation would have to cease while the case was being heard by Commonwealth Court. Proponents, however, expressed little concern because they said there were no grounds for the appeal.

ROBBI celebrated the second anniversary of its existence at a local Howard Johnson's Motor Lodge in February 1988. There was an informal party and a slide show outlining the group's history and listening to words of encouragement from outside activists, who suggested that "the state Senate passage last summer of a mandatory trash recycling bill is a sign of the progress that anti-incinerator forces have made" in recent years (Davis, 1988a).

Disposal of the ash produced by the county incinerator was discussed in some detail at a March 1988 meeting of the York County Solid Waste and Refuse Authority (YCSWRA), but no decision was reached favoring one or another of the twenty-four options outlined by the authority's primary consultant on the project, Malcolm Pirney Environmental Engineering Inc., of White Plains, New York. These consultants estimated that the plant would generate annually 65,000 tons of ash residue which would have to be dumped somewhere at an estimated $40 to $50 a ton. This ash disposal cost amounted to approximately 25 percent of the incinerator's proposed operating expenses (Davis, 1988b).

ROBBI's appeal to the state Environmental Hearing Board (EHB) had still not been heard a year later. While the group's new attorney was seeking to have the incinerator's permit revoked, the York County Solid Waste and Refuse Authority, in October 1988, asked the EHB to dismiss ROBBI's appeal, on the grounds that the group had not met hearing board deadlines in answering certain questions. After ROBBI filed a response, listing its eight founding members, the state EHB denied the YCSWRA motion in November of the same year because the delay in filing had not hurt either side in this dispute.

Although ROBBI's appeal listed a variety of criticisms including water consumption, ash disposal, and others, the hearing board ruling in the spring of 1989 restricted the scope of the appeal to air quality issues, because those were the only issues mentioned in the initial suit filed in the summer of 1987. If it did little else, the appeal by the York protesters kept the incinerator debate alive in the press during this period, when the national movement was gaining momentum, and a *York Dispatch* editorial on July 19, 1989 raised a number of questions about the trash plant:

> The mega incinerator at Black Bridge seems destined to come on line, despite local citizen group protests and despite the fierce debate that is raging nationwide over the safety of mass burn plants. . . .
>
> Perhaps the most daunting problem is what to do with the toxic ash from the incinerator process. According to the *New York Times*, a recent Environmental Defense Fund report, which analyzed ash-toxicity reports from dozens of incinerators around the country, found that bottom ash—the ash that falls through the burning grate—exceeded federal environmental standards for lead and cadmium almost a third of the time. Fly ash—or the ash trapped by air pollution devices in the flue—exceeded the limits more than 95 percent of the time.
>
> Some experts are recommending that the ash exceeding the limits be handled as hazardous waste and trucked to federally-approved hazardous waste sites.
>
> Ash from the Black Bridge incinerator, however, will probably end up with Modern Landfill, located in Windsor and Lower Windsor Townships. So far, nobody seems to be asking how toxic that ash will be.
>
> And the county Solid Waste Authority is pushing the state to dismiss a citizen group's challenge to the air quality permit for the incinerator.
>
> Indeed, it seems that the incinerator project will move forward—no matter what.

The state hearing board considered the ROBBI case only weeks before the new incinerator was scheduled for its first test-burning of garbage in the fall of 1989, more than two years after it was initially filed. The three expert witnesses called by ROBBI were Paul Connett; Thomas Webster, the scientist at Queens College in New York City who worked there with Professor Barry Commoner; and Thomas Germine, an engineer from Morristown, New Jersey. The first two of these witnesses were identified in one newspaper article as having done "extensive research on the health risk assessments for incinerators," and Germine was noted as having taken part "in the building of an incinerator plant" (McCormack, 1989).

ROBBI lost its case, and fewer than a dozen sympathizers showed up in response to the group's appeal to join its members in a two-hour protest after officials scheduled a test-burning in mid-October 1989. Although the group continued to serve as a channel of local protest about noise coming from the operating incinerator as well as about incinerator ash during 1990 and beyond, our account ends at this point because the plant was built and began operation.

Questionnaire responses from nine of the thirteen activists on ROBBI's list provided additional insight into the activities and perspectives of these York challengers. To an open-ended question asking respondents to "briefly describe the type of work you usually did for the group," for example, responses supported the notion that the opposition to the York incinerator focused almost exclusively on litigation. The most commonly mentioned activities were fund-raising and attending meetings. Endorsing this perception by others, the leader described her work: "I am the president of the group. Work consists of planning monthly public meetings for three-and-a-half years and leading the group in fund-raising projects, and keeping updated on important data pertinent to the group." In answer to a question asking about "turning points in the group's development," most ROBBI responses mentioned interactions with attorneys or specific litigation—for example, "When Attorney Lohbauer spoke here in York" and "When Malcom Pierny and the Solid Waste Authority wanted to negotiate with us." Similarly, in response to another item asking about "decisions by the group or its leaders which turned out to be very important," ROBBI activists again emphasized litigation strategies: "Hiring of our present attorney," "When the group voted we needed an attorney," "The decision to contact Attorney Mark Lohbauer." A more critical response to these same items might emphasize ROBBI's decision not to frame the incineration struggle in more adversarial terms or to adopt more militant strategies and tactics. Lackawanna County activists, on the other hand, answered such items very differently.

The Lackawanna County Siting Attempt

In striking contrast to the protracted siting process stretching over several years in York County, the attempt to site an incinerator in Lackawanna

County was defeated within months by widespread citizen mobilization.[7] The first public mention of this proposed incinerator was the cue for citizen action in the spring of 1988, and the project had been effectively stopped by autumn of the same year. This citizen challenge is remarkable because it took place in an area that until that time had become the major dumping ground of the Northeast—without any significant organized protests emerging among local residents against such previous exploitation. Because of this past history, incinerator proponents never expected to encounter such stiff grassroots resistance.

Unlike the other seven siting ventures we are considering, this one was officially sponsored by a local businessman, Louis DeNaples, rather than by county authorities—although it was commonly understood that most county authorities were prepared to go along with the plans.[8] There was no recycling program in Lackawanna County before the start of this project, which was targeted for Dunmore, a small municipality near Scranton, Pennsylvania, adjacent to a landfill also owned by DeNaples. Citizen opposition leaders perceived Dunmore Borough Council President Leonard Verrastro as "basically a nice guy from the old school who saw the plant as a financial benefit for the town and did not worry about its health effects."[9] One of the main reasons these challengers were so successful, according to their own assessment, "was because we got in on the ground floor. Once these things get their permits it's really almost too late for people to start screaming and yelling. That's a problem that a lot of citizen groups have: they're asleep at the switch, and when they finally wake up it's too late." First publicly announced in the latter part of 1987, this project "was all gone by December 1988," according to citizen activist and attorney Diane Beemer, who helped lead the challenge.

A newspaper account in early 1988 portrayed the dispute as one between "the Sierra Club and its supporters," on the one side, and "two companies that plan to construct a 'Mass Burn Plant' on the Keystone Landfill property in Dunmore," on the other:

7. The following individuals were interviewed by telephone or in person in preparing this account: Attorney Diane Beemer, Kevin McDonald, David Byman, Dunmore Mayor Joe Domnick, Josephine Marchese, Sandy Lamanna, John Murphy of the *Scranton Times,* and Bob Curran of the *Scrantonian Tribune.*

8. Neither Louis DeNaples nor anyone from his office would answer questions about this project.

9. Interview with activist Attorney Diane Beemer, Clarks Summit, Pennsylvania, September 24, 1990. Another local activist who helped establish this Lackawanna County grassroots protest group, Kevin McDonald, joined us after an hour or so and shared many of his own valuable insights as well as documents relevant to this dispute.

The war of words over a planned $120 million garbage incinerator continued Wednesday, with opponents claiming it was a "biological time bomb." Supporters steadfastly denied this as a "false claim with no data to support it."

Opponents of the plant have a long list of complaints and assert that the facility presents potential serious health hazards. But the two companies dispute all this and say the plant will be completely safe and either exceed or be in compliance with federal and state safety regulations.

Ground breaking is scheduled to begin next year, with the plant expected to be in operation in 1991. Butcher and Singer Inc., an investment firm with headquarters in Philadelphia and an office in Scranton, and the worldwide Bechtel company would build, own and operate the plant on 10 to 20 acres near the Keystone Landfill.

The Northeastern Group of the Sierra Club in Scranton strongly opposes the construction of the incinerator and is asking its members to complain to their elected officials on the municipal and county levels. . . .

Cliff G. Mumm, Bechtel's project manager in Vienna, VA . . . said there will be no smog . . . [and that] studies have already taken place and they show no chance of smog. . . . Mumm said the [opponents'] statements about cancer and health problems were "scare tactics." . . . Mumm said that there's no evidence that anything harmful can get in the [county's] water supply because the inert material will be buried in a new double-lined landfill at Keystone, and there's also protection from a double pipe.

Nearly all of the municipalities in Lackawanna County have said they'll use the incinerator, Mumm said. Regarding the landfill, he said everything that goes into it will be non-hazardous and that hazardous materials will go to landfills that take such substances.

The plant, Mumm said, will be similar to the company's facility in Cape Cod, Mass., where 60 persons are employed. Bechtel says the plant will be safe, well managed, and good for the community, particularly because there's a problem in the nation, state and in Northeastern Pennsylvania with garbage disposal.

The Sierra Club believes there's a better way and that an alternative is a combination of source separation, recycling and toxic removal that purportedly is a long-term solution to the garbage

> problem. Sierra representatives claim this is cheaper and safer than massburn incineration. (Curran, 1988a)

The Sierra Club thus provided the initial organizational challenge to this proposed incinerator, and it continued to work with the grassroots organizers throughout the conflict. As protest against the incinerator emerged in Lackawanna County, various local politicians spoke out against it. Councilman Steve Davitt of Dickson City, for example, forced his borough council to consider the issue on March 2, 1988, and came away with a unanimous agreement by the four members in attendance to send a letter to the Department of Environmental Resources indicating their opposition to the incinerator. Noting that only Scranton and Clarks Summit had supported the incinerator, Davitt charged Scranton with trying to get "a free ride, while they dump in our back yard, near our schools and plants." A wake-up call was sounded for the people of Dickson City and, indirectly, the rest of the county:

> Noting that Clarks Summit Borough Council unanimously adopted a motion favoring Keystone's expansion plans, Councilman Joseph Prorock facetiously declared that Dickson City ought to adopt a motion to have the landfill located in Clarks Summit.
>
> "Even in Haiti they don't want incinerator ash," Davitt shouted, "but there are people who say 'Give it to the coal crackers because they don't care what you dump on them.' Let's start fighting for ourselves. Our people will have to wake up."
>
> Former Councilwoman Freeda Jezieski opined that the people of the Mid-Valley should get together and discuss this matter. "Have a public meeting," she declared.
>
> Lackawanna County's alleged silence on the issue came in for heavy criticism from several persons at the session. (Lukowski, 1988)

Until an anti-incineration group of local citizens emerged, the Sierra Club of Northeastern Pennsylvania kept the protest alive in Lackawanna County through informational meetings. The major early meeting took place in May 1988 when the Sierra Club brought approximately 100 people together, including environmentalists and politicians, at a downtown Scranton hotel. One of the invited speakers was Paul Connett, who warned his audience that mass-burn incineration only perpetuates landfilling and is actually the most expensive way of handling garbage. Arguing that recycling is a cheaper and safer alternative to garbage disposal than mass-burn incineration, Connett cited a German study allegedly showing that a combination of mandatory source separation, recycling, composting, and toxic removal is cheaper, safer, and just as effective in reducing waste to be landfilled as mass-burn incineration.[10] Attorney Mark Lohbauer, who took over ROBBI's case in York County, also encouraged Sierra Club members to lead the charge in Lackawanna County, educating the people in the debate between recycling and mass-burn incineration.

Two of those in attendance at this May meeting—Kevin McDonald and Edward Flanagan—started the local protest group that was forming at this time. Their network remained nameless for a few months as McDonald, Flanagan, and others worked to expand its support base. Only three months later would they decide upon a name. Among the numerous objections to the project, none was more central than that the county was already the "dumping ground of the Northeast United States." One graph circulated by anti-incineration activists and drawing upon data provided by the Sierra Club compared the 14,000 tons of trash per day disposed of in Lackawanna County with the 450 daily tons generated there. In other words, as the graph made clear, the county itself produced only 3 percent of the "garbage" it was allowing to be stored or burned in the area. The proposed incinerator alone, apart from landfills already in operation, was scheduled to burn 1,000 tons of waste daily—more than twice the amount of garbage generated each day by the entire county. The following excerpt from a letter to the editor by another local resident, who subsequently became an activist in the evolving grassroots protest group, emphasizes the unfairness of this situation:

> Last week I visited a former college classmate from Philadelphia. . . . To an outsider, Northeastern Pennsylvania, our home, has already become "Trashsylvania." . . .
>
> The so-called "Keystone Resource Recovery Project" requires the importation of over 600 tons of trash each day to feed the monster incinerator and make it economical to operate. The operators . . . will make a great deal of money while Scranton remains a large-scale garbage importer. We, the area's residents, suffer health conse-

10. This summary of Connett's remarks combines insights from a telephone interview with Sierra Club officer David Byman, professor of biology at Penn State, September 13, 1990, and newspaper accounts of the meeting.

> quences of dirty air and water while big business makes money at our expense. . . .
>
> We must all recycle today if we are to solve our garbage crisis. (Joyce Hatala, *Scrantonian Tribune*, June 2, 1988)

During the summer of 1988, the Sierra Club worked together with the increasing numbers of grassroots activists to pressure public officials of various county municipalities to address the incinerator issue. Some Jefferson Township residents, for example, cooperated with the Sierra Club to arrange a presentation to the local board of supervisors at their monthly session in early August. Because parts of this township were west of the borough, "uphill but often downwind from the proposed incinerator site near the Keystone Landfill," the impact the project would have was of obvious relevance to its residents. The Sierra Club's local chairman, David Byman, an assistant professor of biology at Penn State's Dunmore Campus, told his listeners at this meeting that "communities in this area need to develop more cooperation in addressing projects and issues which cross municipal boundaries" (Palfrey, 1988).

Proponents of this incinerator project did not stand idly on the sidelines in the face of such increasing grassroots opposition. Bechtel's literature on the incinerator was distributed to elected officials in Lackawanna County municipalities during this period. It included an attractive brochure entitled "The Keystone Resource Recovery Project" which opened out on separate pages to a "Project Summary . . . a Process Flow Chart . . . Benefits to Participating Communities . . . a Project Milestone Schedule, . . . and Development Objectives." Readers were told that the project would "encourage recycling with contract adjustments for current and future recycling programs," offer "employment opportunities—200 to 300 construction jobs; 60 to 70 full-time operation positions"; that permitting was to be accomplished by the first quarter of 1989, when construction was scheduled to begin; and that the project was to be completed by the third quarter of 1991.

Bechtel's packet also included a four-page pamphlet entitled "Ten Common Questions and Answers About the Keystone Resource Recovery Project." In response to the pamphlet's question "Where will the waste supply come from?" the answer given was that the project "will provide waste disposal for communities in Lackawanna County and the neighboring area." The second half of the response was vague regarding the importation of outside wastes for the facility: "The facility does not plan to contract for waste from urban areas outside the immediate region of Northeastern Pennsylvania." Opposition activists insisted that it would be necessary for the facility to accept wastes from New Jersey or New York later, pretending that their original plans had to be revised because of trash shortfalls due to local recycling efforts or other factors.

The incinerator issue occupied center stage in the county during August and September 1988 as municipal officials found themselves challenged to act by increasingly outspoken grassroots opponents. Citizen meetings were being held around the county as anti-incinerator activists prepared a major assault on the project. Some local newspapers gave the opposition more favorable coverage than others. The following article in one of the less sympathetic newspapers reports on an upcoming meeting in the proposed host community of Dunmore in early August 1988:

> Several Dunmore residents are scheduled to meet tonight "setting up an agenda, and trying to get the rest of the borough motivated and moving on this," said Edward Flanagan, who said he's trying to rally residents against the incinerator.
>
> Flanagan said residents would use "public pressure" to convince members of the borough council to vote against the trash plant. . . .
>
> On August 16, officials from Bechtel are set to meet with the council informally on the proposal—which has put the issue in the spotlight. . . .
>
> The Bechtel company had not made a formal application with the borough to build the plant as of last month, officials said, but the company has said in brochures that the trash incinerator would be built next to the Keystone Landfill and that nine counties in Northeastern Pennsylvania would send trash to the plant.
>
> Dunmore Councilman Michael Cummings said the council is "certainly interested" in hearing what residents have to say, but that only "five to ten" people so far have told him that they opposed the incinerator. . . .
>
> The Sierra Club is expected to address the council on the trash plant at a meeting in September. (Lalli, 1988a)

Another area newspaper, the *Scrantonian Tribune,* had been sympathetic to the incinerator opponents from the start, and some activists considered its coverage of opposition activities a contributing factor in the successful mobilization. Its main competitor, the *Scranton Times*, was viewed by project challengers as being in basic sympathy with the incinerator proponents because, as one said, "it is run by a local family which is very pro-business." The *Tribune*, however, experienced increasing financial difficulties during this period and went out of business in the midst of the struggle. Activists viewed this as a serious setback because, in the words of Diane Beemer, "We lost a voice with the demise of the *Scrantonian Tribune*."[11]

In early August 1988, another municipality's officials expressed reluctance to sign a twenty-year commitment to supply trash for the proposed incinerator. Dalton's council members raised various problems with promising that small community's 1,200–1,500 annual tons of trash to the Bechtel project for twenty years. For example, one critical official noted: "There could be major technological advances in five years, let alone 10 or 20, that would cause the borough to regret a two-decade agreement." Another asked what would happen if the facility did not get the required volume of trash, and predicted that it would be trucked in from Philadelphia and New Jersey, "adding to Northeastern Pennsylvania's reputation as a dumping ground" (Emery, 1988a).

By mid-August, Dalton, Jefferson Township, Forest City, and Throop had said no to the proposed incinerator. Sierra Club spokespersons addressed fifty-two questions, widely circulated ahead of time in written form, to Bechtel representatives appearing before the Dunmore Borough Council on August 16. Among them were "If the City of Scranton says no to mass-burn incineration, would your project still be feasible?" and "What will it take for Bechtel to abandon this mass-burn project in Lackawanna County?" This presentation by Bechtel represented the first of two meetings the Dunmore Borough Council scheduled to hear both sides of the issue; the Sierra Club was scheduled to appear and answer questions at its September meeting. The August gathering was hardly a warm one for Bechtel, but it was only a hint of what would happen at the following month's meeting:

> Bechtel walked into a hornet's nest Tuesday night at Dunmore Borough Council's work session while attempting to discuss the merits of its proposed waste incinerator to be built at the Keystone Landfill. . . .
>
> The firm did not grasp the magnitude of its opposition's resolve until it saw about 150 picket sign-carrying people march with lighted

11. Telephone interview with Diane Beemer, June 6, 1994, when she made this and other important points in response to an early draft of this chapter.

> candles in hand to the front of the Dunmore High School singing "America the Beautiful."
>
> It was to be that kind of a night—big business against the little guy. Faced with a hostile audience composed of members of the Sierra Club of Northeastern Pa., the Throop Taxpayers Association, and residents from Clarks Summit Borough, Jefferson Township, Olyphant, Dickson City, Forest City, and members of RESCUE, a local environmental group, Bechtel representatives tried their best to create tranquility.
>
> The hornets stopped buzzing until Jack McNulty, president of the Scranton Building and Construction Trades Council and the Scranton Electricians Union came out dogmatically in favor of the incinerator before an already sweltering audience which was in no mood to discuss economics.
>
> "We have 20 years of experience with Bechtel . . . ," McNulty shouted. "Go, go, go, go, go . . . ," the audience screamed thunderously, stomping their feet.
>
> "Let's get this over with," Leonard Verrastro, council president, pleaded to McNulty. "It's getting hot in here." . . .
>
> Bechtel said Dunmore is one of 90 municipalities in a nine-county region it has spoken to or will speak to concerning the plant. There is "more than enough" trash generated within the nine-county region to support its proposed incinerator without accepting outside trash from Philadelphia, New York, or New Jersey, it said. (Cadden, 1988)

Local opponents had, by now, evolved into a broad-based grassroots organization that would surprise Bechtel representatives and other incinerator proponents at the following month's presentation. Kevin McDonald, one of the leaders of this protest group based in his home community of Dunmore, site of the proposed incinerator, discussed its evolution over the three- or four-month period from May to August 1988 in the previously mentioned interview in Attorney Diane Beemer's office:

> After the Sierra Club meeting [May 1988], Ed Flanagan and I went to an industry information meeting on the incinerator, which was advertised in the paper. We learned more about it, which scared the Bejesus out of us, and then we met with Sharon McAllister and started going door-to-door in the neighborhoods of Dunmore. . . . The Sierra Club helped with their graphs and other data showing the

> big discrepancy between the amount of garbage generated in this area and the amount we take in—something like 20 times what we produce!
>
> Our group didn't have a name at the time, but the three of us went around the neighborhoods all summer showing people the data. . . . Others joined us, and then we had a meeting at my house, and that's when CARE formed—I mean we put a name on our organization—and that was in August of 1988. . . . So, between May and August we were just a bunch of people calling one another on the phone and saying we're going through this neighborhood today, and let's try that one tomorrow, and so on. . . . We couldn't even agree on a name that night in August at my house, but the next morning someone called, suggesting CARE, and we settled on that label for our group.

CARE (Citizens' Alert Regarding the Environment) worked with the Sierra Club to increase the pressure on Bechtel by collecting and publicizing data about problems at its other facilities. The proposed Dunmore incinerator was said to be similar to Bechtel plants in Florida, where there were complaints about water pollution and cost overruns. Citing a *Wall Street Journal* article comparing incineration to the "nuclear power fiasco," one local journalist quoted a Sierra spokesperson as criticizing Bechtel for trying to suppress negative information about one of its own Florida plants:

> [Vito] DiBiasi [Sierra's solid waste chairperson] said: "Bechtel wouldn't give us the information last week before the Dunmore Council and said it's a matter of public record, and that we could get it. Bechtel said it had nothing to do with McKay Bay, but it's in Bechtel's own brochure. This was an eye-opener because we found out Bechtel built McKay Bay." (Curran, 1988b)

The organized opposition was indirectly responsible for the setback Bechtel received in late August, when a state legislator criticized incineration as a trash disposal method and insisted that serious recycling be given priority:

> [P]owerful hometown legislator Sen. Robert Mellow . . . opposes construction of any mass-burn incinerator plant in Northeastern Pennsylvania at present because of an apparent lack of technology and regulations established by the Environmental Protection Agency

> and the shortage of the state Department of Environmental Resource regulations on mass burning plants in the commonwealth.
>
> In a letter to Scranton attorney Michael Cowley, a spokesperson for CARE, which opposes the Bechtel plant, Mellow [said], "I am not convinced at this point it is wise on the part of the Bechtel Corporation to further pursue the possibility of locating their incinerator plant in Lackawanna County or Northeast Pennsylvania." . . . "I strongly believe that the more appropriate position would be to allow the new recycling bill just signed by the governor to be fully implemented so we can determine whether we can, in fact, reduce the amount of waste that we generate in our region through recycling." (Grady, 1988)

The expense-paid trip that vendors customarily offer municipal officials was also deftly used by opponents in early September to undermine the project. A Lackawanna County activist contacted the anti-incineration group in the Massachusetts community where Lackawanna officials were scheduled to view an incinerator built by the same vendor, and this New England protest group requested a meeting with Dunmore officials and others making the trip:

> A citizens' group in Rochester, Mass., which opposed the trash incinerator that Dunmore officials will visit on Monday, said it wants to meet with its Pennsylvania visitors to give them the opponents' side of the story.
>
> "A gentleman [from the Scranton area] called me to tell me that they're coming out Monday, hopefully," said Claire DeLoid, a resident of Carver, Mass., where ash from Bechtel's nearby waste-to-energy garbage plant would be dumped.
>
> Bechtel is giving Dunmore borough council members and interested residents a tour of its just-completed SEMASS incinerator, similar to the one the company wants to build at the Keystone Landfill in Dunmore. . . .
>
> Asked whether she thought Dunmore should approve an incinerator in its own area, DeLoid said, "No. Why do they want it in their town? What is the reason?" . . .
>
> Kevin McDonald of Dunmore, a member of Citizens' Alert Regarding the Environment, said he looked forward to speaking with

> Carver residents about the incinerator. "We'll ask them exactly what the problems are," McDonald said Thursday.
>
> CARE opposes an incinerator in Dunmore because of what it says would be its pollution effects. . . . Bechtel officials have said their incinerator would be designed to protect the environment. (Lalli, 1988b)

Sierra Club officials turned down the invitation to tour Bechtel's Massachusetts plant because they "questioned the motives of Bechtel's all expenses paid junket, especially because the SEMASS plant was not yet fully operational." Dr. David Byman said:

> The club cautions that being wined and dined by Bechtel may become as much of the sales pitch as the actual limited and choreographed display of a non-operational mass-burn incinerator [and] warns that Bechtel's financed trip is typical of junkets to incinerator sites in North America and Europe used by the mass-burn incinerator industry to sway municipal governments. Bechtel likes to show off new equipment, but Bechtel did not provide the Sierra Club with any information on the Bechtel designed and built plant in McKay Bay, Fla., that is presently operational. The Sierra Club has since learned and made public numerous problems at the McKay Bay plant. (Curran, 1988c)

CARE spokespersons requested an inspection of this McKay Bay incinerator but received no response from Bechtel's officials.

The group taking the tour visited the Massachusetts site and was told by Jim Henchel, the Bechtel civil engineer who designed SEMASS and was planning the company's Dunmore incinerator, that its choice was simple: one truck of ash or ten trucks of garbage that the unburned ash represents, because "the ash that remains is about 10 percent of the total volume of garbage that is burned." Henchel also insisted that the ash was at least as safe as, and perhaps even safer than, the garbage from which it was generated:

> As for emissions, he vigorously denied a claim by Kevin McDonald of Dunmore and several other persons representing CARE that the system employed by Bechtel releases dioxin, a carcinogen, into the atmosphere.

> "The dioxin issue is dead," Henchel declared, claiming that the federal government has studied the problem in depth and has concluded that the case against mass burn incineration as a dioxin-producer is closed.
>
> Some persons who toured the plant simply were not so sure, however. (McKenna, 1988)

Scranton's city engineer, John Luciani, told reporters he was reassured by the trip to Bechtel's Massachusetts plant and suggested he would have little problem accepting such a plant in his own backyard. "It was a really educational trip," he said, describing the plant as "beautiful." "Sound-wise, we were outside the plant and couldn't hear noise," he said. "As we approached each component of the plant, it was quite noisy, but from the outside it didn't make any more noise than the brand new Immaculate Conception Church behind my house." Scranton's Mayor David Wenzel viewed a videotape of the site visit made by Luciani and was also enthusiastic. Luciani was candid about his bias in favor of Bechtel before making the trip:

> While Luciani said he went to Rochester with an open mind, he admitted he has long held Bechtel in high regard, a by-product of his undergraduate engineering study at Drexel when that firm was among the most sought-after for jobs. "Bechtel, as far as I'm concerned, has had a number-one rating," he said. "I tried to put that aside, but I knew Bechtel was a quality outfit to begin with and I knew I was going to see something good." (Books, 1988)

CARE established a speakers' bureau in early September, to provide local community groups with its view of the incineration issue. In a prepared statement explaining this move, the group's attorney, Michael Cowley, said:

> Our concern is that the hype which has recently developed will cloud the facts as they relate to this incinerator and its potential effect upon the community of the Lackawanna Valley. To ask our fellow citizens to make personal and financial commitments for a period in excess of twenty years to a mass-burn incinerator owned and operated by outside interests requires that tough questions be asked and forthright answers issued.

Incinerator opponents intensified their efforts as the date for the second Dunmore Borough Council meeting on September 20 approached. Will Collette, an environmental activist from the Citizens Clearinghouse for Hazardous Wastes, based in Arlington, Virginia, came to Lackawanna County on September 14, 1988, to discuss strategies for fighting the plant with local opponents. In a press conference at the Wilkes-Barre–Scranton International Airport upon arriving in the area, Collette said, as reported in a local newspaper:

> "Northeastern Pennsylvania has been identified as 'perfect' by Cerrell Associates, the California firm which specializes in finding areas which are 'least resistant to the siting of hazardous waste dumps.' "
>
> He said the firm looks for areas with the following characteristics: rural, low-income, low education, hungry for jobs, high percentages of elderly, politically conservative, Catholic, and a history of tolerance for land-exploitative occupations such as mining and farming.
>
> Collette said three varieties of "politics" are involved here—partisan (Democrat vs. Republican), back-room (the network of "old boy" insiders), and town-meeting (which is what he says can defeat the back-room boys). He said Bechtel, with strong ties to the Reagan administration, exemplifies the back-room variety.
>
> "We're political in terms of old-fashioned, grass-roots democracy, speaking out . . . , the town-meeting form of politics," he said.
>
> He said he will advise local groups to form a broad base of popular support involving the clergy, children and the elderly. The group will then test the sympathy of elected officials for their cause. (Conroy, 1988)

Bechtel's regional manager, Patricia Sebring, challenged Collette's statements, as reported in the *Scranton Times*: "It's obvious the environmentalists are committed to recycling, because they even recycle their own inaccurate statements regarding the Keystone Landfill Resource Recovery Project." She continued: "Residents of the area should be particularly insulted by the claims that this project was sited based on a very unflattering community profile. . . . The Scranton-Dunmore area could hardly be called 'rural.' " The *Times* also quoted Sebring as saying she was "personally offended" by what Collette had to say about the Bechtel organization (Lyon, 1988a).

A few days before the Dunmore Borough Council meeting, organized opponents received a major assist from the Lackawanna County Medical Society, which publicly urged that there be a moratorium on the construction of new incinerators in the county until further studies were conducted. The local press reported:

> Mass-burn incinerators are a controversial issue in the county and on Tuesday a massive turnout is expected at the Dunmore High School Auditorium at 7:30 p.m. when the Dunmore Council deals with [this issue]. . . .
>
> The Medical Society, meanwhile, said it is "very concerned about the possible environmental ramifications associated with mass-burn incinerators."
>
> Dr. Thomas E. Zukoski, president of the society, said that with the Keystone Landfill area in Dunmore being considered as a site for such an incinerator, "we worry about how that incinerator and its by-products may affect our communities."
>
> The society is asking for a moratorium until studies are conducted by the federal Environmental Protection Agency and the state Department of Environmental Resources on air emission standards for mass-burn incinerators.
>
> The society also wants to know "where and how to get rid of the possible hazardous waste accumulated by ash residue." . . .
>
> "Our garbage crisis may be a cultural crisis, one that may require cultural change in order to stimulate a long term solution. . . . We suggest the mandatory recycling bill recently signed by Gov. Robert Casey be implemented. Then a determination can be made as to the need for a mass-burn incinerator in Lackawanna County." (Curran, 1988d)

CARE activist Attorney Diane Beemer released "The Big Dump List" a few days before the second public hearing in Dunmore. Documenting twelve existing or proposed toxic landfills, incinerators, or dump sites in Lackawanna County, Beemer claimed: "Our region can't survive any more contamination because the ozone levels are high here and we have polluted mine shafts under the entire county." She said that if the proposed mass-burn incinerator became part of the list, "it will be like driving one more nail into the coffin" (Curran, 1988e).

If there was a single critical turning point in this particular conflict, it was the second public hearing at the Dunmore High School on September 20,

1988. As noted above, the first hearing the previous month allowed Bechtel to present its case, even if to a primarily hostile audience. This time it was the challengers' turn, and more than 2,000 residents of Dunmore and surrounding communities gathered to hear speakers criticize the incinerator project. Members of CARE and the Sierra Club handed out pamphlets, sold anti-incinerator T-shirts, and gathered signatures on petitions at the doors. The keynote speaker was Paul Connett, who told his listeners that "more than 200 proposed incinerator projects in the United States have been defeated during the past two years, partly because people believed the democratic process was being subverted by local officials."

> Dr. Connett's main message of the night was that source separation and recycling are alternatives to burning garbage in incinerators, and that people around the world are being bombarded with unhealthy levels of contaminants without the addition of more incinerators.
>
> Incinerators, he said while showing the results of studies in the United States and abroad, have been proven to contaminate the environment with toxic metals like lead and cadmium, and emit cancer-causing dioxins into the air.
>
> "These incinerators are producing some of the most potent cancer-causing agents we have found in the laboratory," he said. "The point is, we're dealing with toxics that are going to be toxics forever."
>
> Dr. Connett also said incinerators are primarily profit-making ventures, accounting for the largest boost in the construction industry since the widespread introduction of nuclear power plants during the 1970s. "What we're looking at here is big business—make no mistake about it," he said.
>
> "The way to go is not to put your faith in magic machines, but back in the people," he said, emphasizing the importance of recycling. "I hope when you leave here tonight, you have your common sense intact and you realize that incineration is the wrong solution to an urgent problem." (Lyon, 1988b)

Kevin McDonald, one of the CARE leaders who helped arrange this event, later commented in an interview on its logistics and outcome:

> Basically, we didn't really have it together until the last minute. We had contacted Paul Connett approximately a month ahead of time to set up a tentative date, and then again the week before our hearing

> to make final plans. We were able to fly him in, but had to drive six hours to get him home. I'll never forget that trip! We couldn't make the airline connections to get him back to work the next day, so we drove him up to his home in St. Lawrence County [New York] after he spoke. . . .
>
> CARE really only became established financially after we collected public contributions at this meeting and asked people to pay membership dues. Before this, during the summer of 1988, I put in three or four hundred dollars for postage, telephone calls, and that kind of thing. . . .
>
> Even though Connett was our keynote speaker, as it turned out the person who was most effective in persuading the audience, from what I could tell afterwards, was Sandy Lamanna, the local woman who gave a very personal account of her experiences in Throop, where, she said, officials of the Marjol Battery Company "tried to shut us up 12 years ago." . . . She pointed to the councilmen and said, "We will hold you responsible. . . ."[12]

Borough officials, although generally reluctant to make public statements on the issue, were obviously influenced by what they heard. Council President Leonard Verrastro, for example, was quoted the next day in the *Scranton Times*, without any byline: "I can't see it coming in right now [referring to the incinerator proposal]. With the emotion of the people, I can't see it coming through. People don't believe in it, and Bechtel officials haven't presented us with an application anyway. So long as they don't apply, it's all over as far as the council is concerned. Personally, after last night's presentation, I don't think any of the council members are going to take a different position. . . . I think it's a dead issue."

Over the next few days newspaper stories quoting "reliable sources" suggested that two scenarios would prevent the incinerator from being built. The first was that the Dunmore Borough Council would reject it if forced to vote, and the second was that Bechtel would drop its plans in light of the overwhelming public opposition revealed at the meeting. A week or so later, an article headlined "Dunmore, DeNaples family agree to not build incinerator at landfill" appeared:

12. Interview with Kevin McDonald in Diane Beemer's law office, Clarks Summit, Pennsylvania, September 24, 1990.

> Announcing that "friendship and mutual concerns about a safe environment were more important than business considerations," the borough and Keystone Landfill have reached an agreement they believe will finally put the incinerator issue to rest.
>
> The heart of the three-page joint statement, written by Dunmore Solicitor Armand E. Olivetti Jr., is that "Keystone will not allow, so long as it is locally owned and operated, an incinerator to be constructed on its lands in Dunmore. Keystone will also agree not to sell off any lands which might be used for the construction of an incinerator until such time as the Borough of Dunmore adopts regulations governing the construction and operation of incinerators." (Emery, 1988b)

Although this issue and what might eventually be built at Keystone Landfill remained in the news for another six months or so, no incinerator was constructed in the area. CARE expanded into a countywide organization that broadened its focus to a variety of environmental problems and in December 1988 had a billboard along a major avenue encouraging passersby to join it to prevent New York, New Jersey, and Philadelphia from "dumping on Lackawanna County." For our purposes, then, 1988 was the end of this successful grassroots challenge.

CARE activists' responses to the questionnaire items about their work, turning points in the struggle, and critical group decisions provide evidence of a much more diverse, broad-based, and politically active group than their York County counterparts. The responses to the open-ended questionnaire item asking "the type of work you usually did for the group," for example, were considerably more varied than ROBBI's. None of the CARE activists mentioned "fund-raising," the most common answer among ROBBI activists. The most frequently mentioned work by CARE activists was "door-to-door canvassing" and "research"; "speeches to local groups" was also mentioned by some. The diversity of the activities of this Lackawanna citizen organization is suggested by some of the other "work" listed: "public relations," "member of board of directors, attended meetings on hazards, made some media contacts," "coordinated efforts of the Sierra Club with the local grassroots," "spokesperson, editor," "handled the money," "organized a phone bank," "contacted other environmental groups and individuals," "develop, print, and distribute flyers, phone calls, minutes at all meetings."

The activists' questionnaire also asked for "turning points in the group's development," and it is not surprising that most of the CARE responses mentioned the September 1988 public meeting. A typical response was: "Community meeting attended by several thousand people where national experts spoke on the dangers of incineration." Other comments reveal a much more dynamic, diversified, and broad-based organization than their York counterpart. For example, one of the founders mentioned as a key turning point for the leaders: "When we grew to six people and had an influx of new ideas and energy." Others mentioned different turning points for CARE: "Sierra Club's small public meeting in May 1988," "The emergence of a public awareness that we had become a sacrifice area," "When we received the support and backing of the Lackawanna Medical Society," and "One local paper's sympathetic coverage—it's now out of business." Finally, one wrote: "I would say the defeat of the proposed incinerator was our key turning point, because with the help of a lot of environmentally conscious people we rallied together and merged as one—to care. And CARE is now a strong and substantial organization."

In response to our question about "decisions by the group or its leaders which turned out to be very important," CARE activists almost uniformly emphasized the organization's decision to reach out to neighborhood groups in the effort to form a countywide environmental organization: "Uniting with other environmental and neighborhood groups," " Traveling to other areas to listen or speak out [on incinerator issues]," "Educating people as to the larger picture of toxins and pollutants in the valley," "Stressing education and door-to-door canvassing," "The decision to stay polite and try to avoid a radical reputation."

Overview and Continuities

The physical and structural characteristics of the York and Lackawanna County projects (Table 5) provided few clues in accounting for their disparate outcomes. Our fieldwork data, however, did reveal two very different processes in each county, which helped account for the results: the gradual emergence of a legalistic, local, and ultimately unsuccessful challenge in York, and a more politically aggressive, broad-based, and successful challenge in Lackawanna. The fieldwork data also provide a useful background for interpreting the backyard survey results (Chapter 3) because they reveal the relative unimportance of most demographic variables and preexisting networks in these cases, and the corresponding importance of widespread grassroots organizing. The much larger percentage of Catholics in the Lackawanna area (74 percent, versus 16 percent in York), for example, was transformed from being the liability the Cerrell Report predicted into a mobilizing asset by creative organizers, thus helping to account for backyarders in that county reporting more negative attitudes toward WTE plants in the telephone survey.

Leaving more systematic comparisons between sitings and defeats until we have analyzed the remaining individual cases, what theoretically interesting insights emerge from the successful sitings in this and the previous chapter? Regarding their actual siting processes, in neither area was there any organized oppositional support from outside the backyard areas. It is also remarkable that, contrary to what some proponents suggest, both Delaware County's and York County's incinerators were built to import out-of-county waste.

And what theoretically relevant generalizations emerge from the two defeated sitings analyzed thus far? Two obvious similarities between the effective grassroots challenges in Cape May and Lackawanna Counties were the presence of splits among political elites in each case, and the widespread involvement of protest assistance from outside the immediate backyard areas. The fact that we also found a split among political elites in the successful Delaware County siting (Chapter 4), however, cautions against assuming too much about the independent importance of that particular variable.

Another question we cannot ignore is how specific cases are influenced by the broader regional, national, and global contexts of the evolving anti-incineration movement. Paul Connett, for example, has been mentioned in three of the four case studies considered thus far—Delaware County being the only exception. And Sierra Club challenges were also important in both York and Lackawanna Counties. Were projects started later in the 1980s doomed because of the organized opposition from Connett, the Sierra Club, and other anti-incineration forces? Some of our data may seem to support such a notion. For example, Delaware County's sited project was first publicly mentioned back in 1982, while Lackawanna County's quickly defeated one became public in 1988. Because both the York siting and the Cape May defeat first emerged as projects in 1985, however, these four cases do not support the hypothesis that grassroots protests emerging later in the 1980s were more likely to be effective because of the increasing strength of the national anti-incineration movement.

6

King Commissioner and Hamstrung Mayor: Montgomery and Philadelphia

Southeastern Pennsylvania is the turf for our third set of conflicts: a successful siting in Montgomery County just outside Philadelphia, and the defeat of a proposed incinerator at South Philadelphia's Naval Shipyard along the Delaware River. These two projects, first publicly announced in the early 1980s, were definitively decided by 1989. Our data were gathered during 1990, while the Montgomery County incinerator was under construction and after the Philadelphia project had been abandoned. The Montgomery County plant officially burned its first municipal solid waste in November 1991—more than three years after Philadelphia's Mayor Wilson Goode, a leading incinerator proponent, publicly announced the discontinuation of his city's waste-to-energy project in July 1988.

Site Characteristics

Montgomery County's incinerator had characteristics some industry consultants would consider likely to favor it over the Philadelphia project as a siting candidate (see Table 6). For example, not only was it approximately

Table 6 Characteristics of proposed incinerator sites in Montgomery County and Philadelphia

	Montgomery Co.	Philadelphia
Size (tons per day)	1,200	2,250
Out-of-county wastes to be accepted?	Yes	No
Estimated population within 2-mile radius	100,000	250,000
Amount of host community annual compensation	$350,000+	0
Citizen advisory committee for incinerator?	No	No
Any previous organized protest in area?	No	Yes

half the size of Philadelphia's proposed plant, but it was also targeted for a less densely populated area that had not experienced any significant previous protest. Perhaps most important is that Montgomery County promised its host community compensation for accepting such a plant, while Philadelphia did not. No special compensation could be promised Philadelphia's backyard residents because no further subdivisions are possible within a city-county such as Philadelphia, whereas various inducements in addition to a $350,000 annual fee was assured the host community in Montgomery County. Neither of these particular siting processes included a citizen advisory committee, and the only variable that some might view as making Montgomery County a more difficult siting than Philadelphia was that the former was planning to accept outside wastes.

This is the first comparison of site characteristics that provide evidence in support of the incinerator project that was actually sited. Because the same factors tended to predict incorrectly (Chapter 4) and were virtually identical in Chapter 5 despite differences in siting outcomes, their real usefulness remains problematic. In any case, we shall see that the complex dynamics of both the siting and the defeat in this chapter cannot be adequately accounted for by these factors.

The Montgomery County Siting Process

This project was first publicly discussed in 1984, but organized citizen opposition within the host township of Plymouth did not emerge until a year or so later.[1] By the mid-1980s, state environmental officials in Pennsylvania

1. The following individuals provided numerous insights that were useful in reconstructing this process: Paul Mack, Mark Lohbauer, Jerry Balter, Jane Garbacz, Bruce Eckel, Don Silverson, Betty Shope, Lilian Davis, and Marie Radatti. Quotations from individual interviews are identified by date in separate footnotes.

had shut down between 1,200 and 1,500 substandard dumps, and Governor Richard Thornburgh increasingly supported the waste-to-energy alternative. Montgomery County itself was said to be planning to build five or six such plants. "The Thornburgh administration advocated mass-burn technology during that time. They called it 'resource recovery,' but that was a misnomer," observed Marie Radatti, an activist opponent from the Plymouth area.[2]

The Plymouth Township Council was in the process of considering establishing a "resource recovery district" and a Montgomery County incinerator proposal in mid-January 1985.

> An incinerator large enough to burn trash from 11, and possibly 23, municipalities in eastern Montgomery County could be built in Plymouth Township under a proposed ordinance before the Township Council. . . . The ordinance would establish a resource recovery district and would allow construction within the district's boundaries of an incinerator capable of burning up to 600 tons of trash a day. . . .
>
> [Council President Robert D.] Wurzbach said Montgomery County had proposed building an incinerator capable of burning up to 1,800 tons of trash a day [at another site]. . . . Wurzbach responded to the county's incinerator proposal by declaring that council would "not allow anybody to turn Plymouth Township into the trash capital of the world." . . . At Monday's meeting, Wurzbach acknowledged that the proposed ordinance might be interpreted as a change of heart by council. . . .
>
> "Up to now, council has been against the county's incinerator proposal, [but] . . . the solid waste issue is hitting local municipalities drastically in terms of money. We're paying $10.37 a ton now to dispose of trash in a landfill. We could pay $35 to $38 very soon, and $50 to $70 a ton is a possibility."
>
> Wurzbach said the resource recovery district would enable the township to control "what goes in, the size and number of trucks, the roads they use. It's better than having some judge tell us we have to build an incinerator to handle 2,500 tons of trash a day." Wurzbach said he was not sure if Plymouth's council could act unilaterally

2. Telephone interview with Plymouth opposition leader Marie Radatti, December 21, 1990.

> to establish guidelines and size restrictions for incinerators. He said the issue might have to be settled in court.
>
> "We in Plymouth Township have to be responsible," Wurzbach said. "Everybody can't say, 'Not in my back yard,' or trash will pile up in everybody's back yards." Township officials said they had not decided when they would vote on adoption of the proposed ordinance. (Pfitzer, 1985)

A few weeks later, a county official said he expected to receive cooperation from Plymouth Township in their preparations for siting an incinerator. The chairman of the county commissioners told a reporter he looked forward to completing such a deal by the end of 1985.

> Montgomery County hopes to have a signed contract by the end of the year with a private firm to build a trash-to-steam facility in Plymouth Township, according to Paul B. Bartle, chairman of the county commissioners.
>
> A management team is preparing bid specifications for private vendors to build a plant that will be able to handle some 1,000 tons of trash daily. . . . Prior to yesterday's commissioners' meeting, Bartle said he talked with the state Department of Commerce which said it would provide the county with more than $100 million in industrial revenue bonds to pay for the project. (Gibbons, 1985a)

At the February 1985 meeting of the Plymouth Township Council, some local residents raised questions about the proposed plant. Opposition would continue to increase over the following months.

In April 1985, Plymouth's council voted to approve the first resource recovery district in the country—an area especially reserved for such things as incinerators and trash transfer stations—despite growing criticism from local residents. Only by the time of this critical meeting was the grassroots opposition beginning to become an organized and vocal presence.

> Members of the Plymouth Township Council withstood four hours of barbs and challenges from about 150 incensed residents and finally approved what Council Chairman Robert D. Wurzbach called the nation's first resource-recovery district.
>
> By a 4–1 vote, the council created a special zone where only trash-to-steam plants or trash-transfer stations may be built. . . .

> For the first time in nine months of meetings, disgruntled residents packed the town hall auditorium Monday night, and they strongly objected to the council's solution to the area's trash crisis. . . .
>
> "This is my life you are talking about, my breathing, my health," said Harriet Savitz. "To hell with the garbage and the tax money. I want to send out a message to Montgomery County that this little township is going to say 'No.' "
>
> Most of the residents who attended the meeting conceded that their criticisms were coming after the council had held five public hearings and already had given preliminary approval to the district.
>
> "It doesn't matter that we are late," Savitz said. "We have been apathetic and wrapped up in our own private business. But somehow, we have been awakened, and we realize that a garbage dump may be in our back yard if we don't do something about it" (Manly, 1985a).

Council Chairman Wurzbach said that he feared losing a court battle with the county if the township council refused to create the resource recovery district, but the sole dissenting council member, Joan Grohoski, argued that the council was really unprepared to make such a decision.

Throughout the spring of 1985, discussions between county and township officials over size, health issues, and even the number of incinerators to be considered for Plymouth Township continued. In April, shortly after the above public meeting, the protest organization TRASH formed.

> Township residents angry over Plymouth Council's decision Monday night to create a new zoning district for trash disposal have decided to fight trash with T.R.A.S.H.
>
> Township Residents Against Scandalous Hazards (TRASH) was born last night in the rustic Kings Road home of Harriet May Savitz, where about 20 residents gathered to plan a protest Monday evening opposing the new district's establishment.
>
> "We want to do something that will let people know that we're not for what the commissioners have done," Mrs. Savitz said, referring to the former titles by which township Council members were recognized in the 1950s and 1960s. "We're protesting those township commissioners and the Montgomery County commissioners who threw Plymouth Township down the drain." . . .
>
> Contacted before the meeting yesterday, Mrs. Savitz explained she

> was concerned about potential health hazards and a decline in property values should an incinerator be built in the new district. . . . "There's no district like this in the country," she continued. "There's nowhere you can look to find out what happened to other people subjected to this type of thing." . . .
>
> Many residents at last night's meeting admitted they did not know about the proposed district or its ramifications until Mrs. Savitz telephoned them yesterday to recruit supporters for her campaign. (Purifico, 1985a)

While individual residents were being recruited from the grassroots for this protest, another level of organizing was being carried out by the leading proponent of the project. Montgomery County Commissioners Chairman Paul Bartle Jr. was coordinating municipalities promising to send their trash to the proposed Plymouth incinerator.

> Half of the 22 municipalities in eastern Montgomery County have signed letters of intent to allow the county to control the disposal of the region's trash. County Commissioners Chairman Paul B. Bartle Jr. said Thursday that the letters from the municipalities, sent at the county's request, were not legally binding but would serve as an indicator of how much waste could be guaranteed if the county built a trash-to-steam plant in Plymouth Township.
>
> Bartle said the letters further provided a measure of how much municipal support there was for the county "to take a leadership role in solving the area's trash crisis." (Manly, 1985b)

At one of the first organized protests by local residents, approximately eighty people marched and waved signs in front of the Plymouth Township Building in late April 1985. Two people who spoke at this early demonstration were Democratic council candidate Norman A. Klinger and Keith A. Kopach, a TRASH member:

> Kopach and Klinger urged residents to attend Council's May 13 meeting, at which it will vote on boundary lines for the new zoning district. "They [Council] are depending on our apathy and ignorance," Kopach told the crowd.
>
> "Somewhere along the line [Township Council President] Bob Wurzbach lost his point of reference as a representative of Plymouth

> Township and tried to solve the problems of [Montgomery County Solid Waste] District 2," Klinger said, referring to a plan to divide the county into districts to facilitate waste disposal operations. "If he wants to be a Good Samaritan, that's great. But he should leave office," Klinger said. (McCaffrey, 1985a)

Starting on April 24, Plymouth residents opposing the township council's creation of a zoning district for trash disposal scheduled four-hour daily vigils outside the township building until the council's May meeting.

Because the county owned the land in Plymouth Township, County Commissioner Bartle appeared determined to build an incinerator there, regardless of the wishes of the local people, to handle what many regarded as a countywide disposal crisis. Some TRASH activists themselves agreed that it was virtually impossible to prevent a plant from being built. Because there was also a possibility that more than one incinerator might be constructed on the newly zoned property, these opponents suggested that limiting it to only one plant might be considered a partial success.

> [TRASH member] Klinger agreed that the county can build a plant if it wishes, but suggested trying to keep it out rather than accepting it without a fight. . . . He suggested that the council keep the regulations which it approved for the district last week, but prohibit trash incineration. "If you're totally successful, you'll have no incinerator," he said. "If you're partially successful you'll get one plant." . . .
>
> [TRASH member] Ms. Savitz said . . . that one plant will be more than enough to ruin the township. "When you have 1,000 trucks [a week], your roads are gone, your air is gone," she said. (McCaffrey, 1985b)

In response to Bartle's accusation that Klinger was orchestrating TRASH to further his own political ambitions, TRASH member Keith Kopach said in a prepared statement:

> Paul Bartle is deluding himself if he believes the current trash revolt in Plymouth is politically motivated. . . . Concerns for the environment and the health of the community are not partisan issues nor should they be. We are a group of citizens that have said to our representatives they have stepped over the line and we've had enough.

> We are fighting against the arrogant, insensitive mindset that allows politicians to tell us we have no choice about what they want to force on our community. No one party or affiliation monopolizes these concerns.
>
> Considering the political strong-arm tactics Bartle has employed against municipalities, it is ironic that he should accuse anyone of playing politics. We invite Paul Bartle to the Plymouth Council meeting on May 13 so that he may clarify his views on the trash crisis and its effect on the residents of Plymouth. (McCaffrey, 1985c)

Although the resources available to TRASH activists in opposing this project seemed to pale in comparison with those of the county, the Plymouth grassroots protest organization made its presence felt. By late April 1985, TRASH had pressured Plymouth Township officials into more of an adversarial position toward the county. When, for example, Plymouth Township Council President Wurzbach created a public stir by using the word "peanuts" to characterize Bartle's description of Plymouth's 16,500 people compared with the 665,000 in Montgomery County, Bartle denied using the word "peanuts" or any such comparison. Admitting that "peanuts" was his own word rather than Bartle's, Wurzbach said he was bothered that Bartle would not even admit making such a comparison in the presence of other witnesses at an administrative workshop session between the county and the township the previous month:

> Wurzbach denied that Bartle claimed Plymouth Township was "peanuts" compared with the best interests of the county's 665,000 voters. . . . "That's an analogy I made," Wurzbach said. . . . "I did not quote him directly." . . .
>
> However, Wurzbach claims Bartle did say the county was going "to proceed with a resource recovery district on land that it owns" with or without Council's approval. . . .
>
> "He said the county will proceed with its plans to place the unit in Plymouth Township whether we liked it or not." Wurzbach said he and Councilmen Frank Zellner and John J. Washeleski were "very upset" at that meeting that Bartle "was willing to sacrifice our township citizens."
>
> "We didn't feel that was right," Wurzbach said. "We argued vehemently that we have to protect our citizens."
>
> Wurzbach said he was not pleased that Bartle denied any com-

> ments about county-versus-township interests at last month's meeting. "I don't like it," Wurzbach said. "There were three of us there at that meeting, Zellner, Washeleski and myself. I can only say that's what I heard at the meeting." (Purifico, 1985b)

Plymouth Council had to repeal its creation of a new zoning district for trash disposal facilities in late April, after being informed by its township solicitor Arthur Lefkoe that "it had violated state law by not holding a properly advertised hearing to consider 'significant changes' in the ordinance creating the district" (McCaffrey, 1985d). While approximately fifty opponents attended the session for this announcement, because they were contacted by last-minute word-of-mouth and telephone tips, TRASH activist and Democratic candidate for council Klinger criticized township officials for not publicizing the meeting so that more people could have been there. Some activists had peanuts taped to their lapels in mocking reminder of Bartle's alleged remarks.

TRASH members organized opposition rallies at the Plymouth Municipal Building, at the Montgomery County Court House, and elsewhere during May 1985, in efforts to pressure decision-makers and increase their own base of protest support. The following letter to the editor from TRASH activist Marie Radatti appeared in the Sunday edition of a local paper, *The Recorder*, on May 12, 1985:

> This is an open letter to my neighbors regarding the proposed trash-to-steam plants in Plymouth Township. I want to address those people who are quietly waiting until Plymouth Township and Montgomery County decide their fate for them.
>
> The clock is winding down, and when you finally decide to make a move, it may be too late. All the decisions will have been legally made and you will have to live with them. The pollutants won't stop at the Plymouth boundary lines. They will also fall on Conshohocken, Whitemarsh, and surrounding areas.
>
> Involvement time is today, not tomorrow, for tomorrows never come.
>
> For more information, or to do something about these vital decisions, write now to TRASH, P.O. Box 181, Plymouth Meeting PA 19462.

Radatti and other TRASH activists were concerned that rezoning would not only open the door to the county incinerator but also invite others to construct such facilities on the rezoned land. Plymouth real-estate dealer Charles Tornetta, for example, was also proposing to build two such plants at this time. In an attempt to dispel the mounting opposition, he took forty-seven people, including both opponents and supporters, to Baltimore for an inspection of a recently completed trash-to-steam plant. He also arranged a luncheon with reporters to discuss his proposal with two engineers favoring the technology.

> Tornetta's campaign comes as the pressure builds on Plymouth officials to decide between Tornetta's proposal and that of Paul B. Bartle, . . . who is seeking a private company to build a similar plant on county land in Plymouth.
>
> Bartle said recently that he intended to proceed with the county's plan, despite Tornetta's pending application before the township zoning hearing board and the fears of Plymouth residents that two plants might be built. . . .
>
> A key vote is expected May 20, when the zoning hearing board votes, in effect, whether to allow trash-to-steam plants in the township. . . .
>
> Tornetta and his three brothers, who are partners in the Fourtrees Co., own about 165 acres near the Alan Wood Steel Plant and have proposed leasing 25 acres to American Ref-Fuel Co. to build a plant capable of handling the 1,200 tons of trash generated daily by the 22 municipalities in eastern Montgomery County. (Manly, 1985c)

After going along on Tornetta's trip, Marie Radatti said that what she saw made her opposition to such a facility in Plymouth more resolute. With more than 100 TRASH members and others in attendance on May 20, Plymouth's zoning hearing board rejected Tornetta's proposal. The activists interpreted this as a minor victory in their struggle, although the decision had no bearing on the plant proposed by Bartle and the county.

By the end of May 1985, TRASH spokespersons had come to accept the likelihood that there would be at least one incinerator built on the county-owned land in Plymouth, and they were trying to keep it as small as possible. After a visit by state legislators to a special TRASH meeting, a group representative told a reporter the activists were looking for "some positive reaction from the legislators about creating a smaller 200-ton-a-day facility which the group sees as feasible for Plymouth Township and surrounding municipalities."

> This is the first time the group has acknowledged it would support a trash facility in the township. Council has held three hearings the past two months trying to establish a special district regulating trash disposal should such a facility be built.
>
> "We have come to realize that we could have a 100-ton-a-day to 200-ton-a-day burner in Plymouth Township that would take care of Plymouth, Norristown, Conshohocken, Whitemarsh, and maybe even a fifth municipality," [Harriet] Savitz said. (Purifico, 1985c)

Rather than agreeing to attend further hearings or to make a public statement on behalf of TRASH, however, these legislators "were trying to make us understand that we would probably get a 1,500-ton-a-day facility," Harriet Savitz went on to explain. "They said the best they could do was tightening up regulations governing the bill Bartle has in the state," she said.

This debate went on into the summer of 1985, with the stance of Plymouth Township officials toward the project becoming increasingly influenced by the TRASH group's opposition to Commissioner Bartle's proposals. A major problem with unqualified opposition to the county's plans by Plymouth Township, however, was the likelihood that the courts would support the county in any legal battle over this issue and that Plymouth would lose all opportunity to control the number and type of facilities imposed upon it. Officials in surrounding municipalities remained uninvolved—nearby Conshohocken, for example, was unwilling to speak out against the plant even though its "odors, fly ash, and trucks will come to Conshohocken," as one of its local officials put it. Reasoning that neither Plymouth nor Conshohocken could stop the county from building a plant, he warned his constituents:

> "You're going to see something—there's a trash crisis in the county." . . . Pointing out that Plymouth has proposed charging a 25- to 50-cent user fee per ton of trash dumped, [Councilman Charles Kelly] said, "Plymouth is going to reap the harvest and we're going to have the problem." (McCaffrey, 1985e)

Throughout the summer of 1985, TRASH activists worked to expand their support base, Plymouth officials adopted public positions increasingly opposed to any plan for a large incinerator, and supporters of the project sought answers to critics' questions about such issues as dioxin's effects on humans and why Sweden had issued a ban on new construction of trash plants. Increasingly pressured by TRASH activists, the Plymouth Township Council reversed previous decisions and proposed its fourth trash plan in early July 1985, drastically reducing its maximum capacity from 1,500 to 250 tons per day (Manly, 1985d). Days later, the same council passed a new law permitting the construction of no more than one 250-ton-a-day incinerator. During July and August, TRASH activists circulated petitions urging people in nearby municipalities to become involved before they found themselves and their neighborhoods suffering the consequences of dioxin poisoning. A Plymouth zoning board member sent a letter to Commissioner Bartle in mid-July, citing an article by Paul Connett and asking that the county build a recycling plant instead of an incinerator. Norman Klinger, TRASH activist and attorney, wrote a lengthy letter for *The Recorder* in early August, listing recent developments he interpreted as indicating that "the Plymouth Township trash battle appears to be tilting in favor of the citizens." Among the specifics mentioned were the zoning hearing board's ruling against a 1,500 tpd plant and the township council's restrictions on incinerator size to 250 tpd. The letter's main point was that local residents should join with TRASH in monitoring the Plymouth Town Council's activities and decisions.

Some surrounding municipalities sided with the county against Plymouth, apparently viewing the whole conflict in NIMBY perspectives rather than seeing the incinerator as something that would have an impact on communities outside the backyard area. In supporting the county's plan, for example, both West Conshohocken and Upper Merion spokespersons mentioned that others did not help their counties in the past when they had problems. In response, an opinion piece by a TRASH supporter criticized both of these neighboring municipalities because they had never requested outside assistance, whereas Plymouth activists were requesting it at this time.

A second hearing on the proposed county trash plant was scheduled in mid-August by the Montgomery County commissioners, and County Commissioners Chairman Bartle announced ahead of time that "the county will present witnesses to dispute testimony presented at last month's public hearing on the matter."

> Of particular concern to the county commissioners was testimony at last month's hearing by Karen Shapiro, an environmental biologist.

> Shapiro testified about studies that indicate dioxin, the most toxic manmade substance known, is produced when plastic and paper are burned together in trash-to-steam plants.
>
> Bartle said the county will present an expert witness on the possibility of dioxin being generated by the resource recovery facility. The county commissioners anticipate that a major portion of the next hearing will be taken up by the county's own expert witnesses. However, Bartle said, a portion of the meeting will be reserved for public comment and questions. (Gibbons, 1985b)

This second county-sponsored public hearing only intensified the debate between supporters and opponents of the Plymouth incinerator project. One editorial in the *Main Line Times*, which serves municipalities farther removed from the proposed site, favored the county proposal and urged readers to attend the hearing to neutralize local backyarders opposing it. At this August meeting, the same combustion expert mentioned in Chapter 1 who was challenged by Barry Commoner, Paul Connett, and others—Floyd Hasselriis—assured his listeners that "dioxins cannot survive a properly controlled 1,600- to 1,800-degree flame" but that "if you don't burn trash properly you will get smoke—and where there's smoke, there can be dioxin." The county's speaker on dioxin, Alf Fischben, insisted that the health effects of dioxin on humans is very limited. TRASH activist Norman Klinger cross-examined the county's experts on combustion and dioxin for almost an hour after their presentations. Other opponents at the hearing "vigorously contested the experts' testimony, and a doctor, an engineer, a retired chemist and a statistician from the audience disputed the validity of studies cited by the county's witnesses" (Brooks, 1985). TRASH leaders presented the commissioners with a petition signed by 5,662 plant opponents from Plymouth and two nearby municipalities. A few days later, Klinger published an appeal for the Plymouth Township Council to establish a "Blue-Ribbon Panel" to help it address the incinerator issue, and the League of Women Voters strongly supported this request, telling readers that Bartle opposed any citizen advisory panel "until a facility is constructed." Whatever name it is given, League officials had in mind a citizen advisory committee (CAC), which consultants for such projects suggest setting up before any site is selected.

TRASH activists continued to focus upon mobilizing local residents, while Commissioner Bartle and other proponents concentrated their efforts on obtaining the support of municipal officials throughout Montgomery

County for the Plymouth incinerator. TRASH leaders, for example, modified the group's name in September 1985 to make the "S" refer to "Solid Waste" rather than "Scandalous," in an appeal to potential conservative supporters who objected to the shrillness of "scandalous." By late August, several municipalities within Montgomery County had agreed, in writing or orally, with the county's plan, and five proposals for construction of the facility had been received. Such facilities were increasingly being favored at both the federal and state levels by the Environmental Protection Agency and Pennsylvania's DER, respectively.

In the midst of accusations that Bartle was involved in conflicts of interest affecting political contributions from business associates of a firm competing to build the Plymouth plant, that firm was eliminated from the short list of three finalists selected for consideration in early September. The only Democratic commissioner, Rita Banning, criticized Bartle and called for the formation of a CAC, or what she referred to as "an independent citizens' panel to review proposals," because it was the only way anything can be salvaged of the county's tainted selection process. No such panel was created, however, and the following month the Dravo Corporation of Pittsburgh was selected by the county to build the plant, which county officials hoped would be operating by mid-1989. In October, Commissioner Bartle announced more restrictive rules regarding audience participation at Montgomery County meetings, after the Plymouth Township Council followed the TRASH lead in officially opposing the proposed plant and opponents staged a disruptive protest. Henceforth, according to these rules, people who wanted to speak at county commission meetings were required to register with the secretary upon arrival and were limited to three minutes of comments.

Protesting such arbitrary restrictions on audience participation at county meetings, a letter to the editor in early October 1985 by one of TRASH's founders, Harriet May Savitz, related the trash struggle to larger issues of democratic decision-making and provides a glimpse at the evolution of the local struggle and activists' perspectives on events:

> I find myself asking "What am I really fighting for?" Is it a battle against plants—or rather a desperate cry for participatory democracy?
>
> There is nothing that can match the horror of watching one's right to participate being taken away. Week after week at Montgomery County meetings, I heard good, intelligent people speak out and beg

> for solutions that they could live with. Week after week, I saw their frustration as they left the room.
>
> It's really democracy that's the issue and the cry I heard was the pain of its illness. On the trash-to-steam issue, democracy would have thrived had a committee of citizens from Plymouth Township and surrounding municipalities been brought in from the beginning and asked to research, to explore recycling, trash-to-steam and other types of incineration. Those who must live with the results of these decisions must have a say in those results. Is that not a democracy?
>
> I believe in the will of the majority. If I do not like what the majority tells me, then it is up to me to leave the area . . . But can a democracy work when just three commissioners decide the future of over 700,000 without the majority of those 700,000 having an input? Is it a democracy when the public reacts, attends meetings, studies, spends its money, begs for alternatives, then waits in fear for the decision?
>
> The people came to County meetings quietly at first, but the rumble now heard is one of frustration and anger. The anger, I believe, is healthy, for it shows that democracy is alive and fighting to survive, and when that democracy is in danger, "we the people," as so eloquently put many years ago, remember that it is up to us to preserve it.
>
> It is not the Commissioners I am fighting against, but democracy I'm fighting for. I want to be heard, but more important, I want that voice to be respected, for it's part of the voice of the people and that sound must not be stifled. . . . (*The Recorder*, October 6, 1985)

By this time, TRASH activists had pressured local elected officials to support their challenge. Tensions between county proponents of the incinerator, on the one hand, and TRASH with its official supporters in Plymouth Township, on the other, continued throughout the rest of 1985 and into 1986. Five county representatives took the customary expenses-paid trip to a distant plant—this time to Germany in late October—to view a comparable modern incinerator. Meanwhile, county officials were pressuring municipalities to sign agreements to send their trash to the proposed incinerator, and Montgomery County commissioners supported Delaware County's case at the State Supreme Court level, arguing that the county should have the right to override local ordinances to the contrary in siting such plants, a case that had obvious relevance to its own situation. On December 11, 1985, the county commissioners signed the necessary documents for the construction and operation of a 1,200-ton-per-day incinerator in Plymouth Township by the Dravo Corporation, with the provision that the county begin construction by March 1988 or lose its bond issue. Of the county's twenty-four municipalities, fifteen had signed contracts promising their trash to the facility by the end of 1985, one had conditionally agreed to do so, and eight had refused (Gibbons, 1985c).

During the same period, TRASH activists worked with Plymouth Township officials to thwart the county's plans. At the heart of the township's argument against the county was the township zoning law allowing one trash-to-steam plant with a capacity of no more than 250 tons a day. Commissioner Bartle, however, insisted that the county had the right to build whatever size plant it wanted in Plymouth. He said: "The township cannot limit the size of the plant; that is an administrative question rather than a zoning issue. . . . It's like telling a grocer that he can build a store but can't sell milk" (Manly, 1985e). At a late November meeting of the county commissioners, Harriet Savitz accused Bartle of "running a dictatorship" because he was "intimidating and blackmailing" municipalities into signing contracts agreeing they would bring their trash to the Plymouth plant (Gibbons, 1985d). Various municipal leaders publicly echoed Savitz's criticisms. By this time, TRASH activists were writing officials in other municipalities, urging them to refuse to sign agreements sending their trash to Plymouth, while also working with their own local representatives in filing legal action to stop the incinerator. TRASH sponsored rallies and mobilized groups of citizens to attend various municipal meetings where they pressured officials along these lines.

In January 1986, the Montgomery County Commissioners adopted a 300-page waste-management plan for all the county's sixty-two municipalities that included the construction of the controversial waste-to-energy plant in Plymouth Township. Although the county had already signed a contract with the Dravo Corporation to build the plant, this was a formalization of county policy and a response to the Plymouth Township challenge. TRASH activist Savitz, however, told a reporter a few days later: "I don't believe that plant will ever be built, but if it is, at least we've put the politicians on notice that the public is not asleep" (Wiegand, 1986). The county responded to the lawsuits by TRASH and Plymouth Township in early 1986 by filing a counterclaim, asking the court to prevent the township from delaying the construction of a 1,200 ton-per-day plant. In response to suggestions that he might consider allowing nearby Philadelphia to dump some of its trash in Plymouth, Commissioner Bartle publicly declared any such use of the new plant "a dead issue" and assured the public "that no Philadelphia trash will be dumped at the facility" (Gailey, 1986). Because some citizens began demonstrating against the Plymouth plant outside Bartle's home in the evening, a six-member courthouse security department was formed in January to guard the county commissioners' offices.

When a grassroots protest relies primarily upon litigation, the odds against a successful challenge usually increase, because the courts are seldom responsive to such groups' negligible political power. Throughout 1986, this particular struggle shifted even more significantly in the county's favor, however. While awaiting the outcome of the court battle with Plymouth Township, Commissioner Bartle was also lobbying the Pennsylvania State Senate to change Act 97, which gave municipalities the authority to solve their own individual trash-disposal problems, so that "the county shall be the entity to develop comprehensive solid-waste management plans with appropriate reviews and comments by local officials." Although incinerator proponents told of some local DER officials critical of incinerators at this time, top DER administrators were squarely in Bartle's corner, supporting trash-to-steam technology. In early April, for example, DER Secretary Nicholas DeBenedictis told an audience that the department's goal for 1996 was to have such plants processing 50 percent of the state's solid waste because so many landfills had to close after being found to be improperly designed and operated. Meanwhile, some council members and others in Plymouth Township publicly complained about the legal bills coming due for the township's battle against the county (Pulver, 1986).

By mid-April, many local activists were again criticizing Plymouth council members, who now stood on TRASH's side in the controversy, for being unaware of the state's proposed new emissions standards for trash-burning plants and for not knowing that hearings on a State Senate bill allowing the county to tell its municipalities where to burn trash would close within a week or so. In response to the new Township Council President Joan Grohoski's excuse that board members are "only part-time employees," a TRASH member responded: "We also work 40 hours a week, yet we know what's going on." Marie Radatti, the new head of TRASH by this time, had arranged for her group to testify before the state hearings ended, and in response to her suggestion at an April council meeting "Grohoski told township solicitor Arthur Lefkoe to ask the state to hold another hearing on the bill in Plymouth" (McCaffrey, 1986).

Although this conflict never focused primarily at the grassroots or township level, by the spring of 1986, it had moved decisively to the courts, regulatory agencies, and the legislature. Marie Radatti, reflecting back on the years of struggle, said their group's situation was quite different from the cases usually addressed by the Citizens Clearinghouse for Hazardous Wastes, the national environmental organization:

> We started out trying to learn from the CCHW suggestions and literature, but our situation quickly passed beyond the kind they were addressing. Montgomery County was using more sophisticated methods than what CCHW envisioned proponents would employ. Bartle was running this county like a king or monarch, forcing municipalities to do what he wanted them to do. We just don't have enough checks and balances to compensate for the power of a person like Bartle in this county.[3]

A Philadelphia attorney who assisted the TRASH activists at various points in their struggle agreed that "Montgomery County, in the person of Paul Bartle, forced the plant on Plymouth residents."[4] It is a moot question whether this challenge might have been more successful, despite such alleged concentration of Montgomery County political power, if it had started at the first public mention of the incinerator project.

Outspending Plymouth Township in legal fees by approximately a 10-to-1 ratio during the first half of 1986 ($634,00 to $67,000 by July), Montgomery County officials were said to have received what some members of the township council referred to as "red carpet treatment" by the DER, the agency responsible for issuing the required preconstruction permits for such things as earth distribution, water discharge, air quality, and solid waste (Gibbons, 1986a). The previous December, Plymouth Township filed a suit against the county asking the Common Pleas Court to issue a declaratory judgment blocking the project, and in August 1986 Lancaster Senior Judge Wilson Bucher ruled against the township, ordering it to "cease and desist from taking any action . . . inconsistent with the expeditious development, construction and operation of the proposed resource recovery facility."

This ruling that the proposed incinerator was not subject to local zoning laws was a major blow to opponents' legal strategies in challenging the

3. Radatti interview.
4. Telephone interview with Philadelphia Attorney Jerry Balter, September 19, 1990.

plant. Judge Bucher had been specially appointed by the State Supreme Court to preside in all litigation between the county and the township over the project. Bartle said the county would immediately contact township officials in hope of reaching an "amicable solution," but TRASH activists—insisting that such secret meetings between county and township officials were really responsible for their problems in the first place—demanded that at least two of their own members also be present at any such meeting "because we truly represent the people's wishes." TRASH attorney William Klinger said, "After all this secrecy that went on for a year and put us in this position today, we cannot afford not to be part of this discussion" (Gibbons, 1986b). Radatti and Klinger issued statements claiming that some township officials had been manipulated by Bartle in the initial stages of the agreement between the county and township. Pressured by these organized citizen opponents, Plymouth Township officials appealed Bucher's decision. A pro-incineration editorial in the *Philadelphia Inquirer*, August 25, 1986, supported the county against the township:

> Absolute veto power over trash disposal facilities by local governments would be a serious impediment to regional solutions. Montgomery County needs to move ahead with its trash-to-steam proposal with all possible speed. If further litigation can be avoided through negotiations with Plymouth Township, so much the better. If not, the county should push for an early decision on appeal.

Because few municipalities stood with Plymouth against Bartle's plan to build the incinerator in its backyard, public hearings by the DER on the issue in October 1986 saw Plymouth Township officials and its citizen group, TRASH, significantly overmatched politically. In addition to the county commissioners, the Pennsylvania Supreme Court and the DER supported the siting of the Plymouth incinerator. Because Montgomery County officials had also announced that Plymouth's appeals would cost taxpayers at least $1 million and that they might also increase tipping fees for users of the facility, neighboring townships had economic incentives to oppose such appeals.

The environmental organization, Greenpeace, pledged its support for the opponents, however, and sent a representative to a press conference at the home of Marie Radatti before the DER public hearing. What was billed as "an informal fact-finding public hearing" turned into a six-hour marathon at a school auditorium in Plymouth, where DER officials heard arguments for and against Dravo Corporation's application to build the plant. Approximately 250 people, with a majority opposed to the project, alternately cheered and booed as experts on each side of the issue spoke to a crowd dotted with dozens of black balloons and black armbands symbolizing "the deaths of children in the future" resulting from dioxins emitted from the plant, according to Radatti. Officials from six of the seventeen Montgomery County municipalities whose trash would be burned in the plant spoke in favor of granting permits, to the audience's cries of "Put it in your township." It is not too surprising that Paul Connett was among those speaking against these permits (Conroy, 1986; Michael Gibbons, 1986a).

The difficulty involved in distinguishing between the claims of scientists on opposing sides in technological controversies was illustrated by the comments of an official in a municipality adjacent to Plymouth. Conshohocken Council President Gerald McTamney said, in a November 1986 interview, that he favored the building of the plant "only if it is safe." After noting the frustration of hearing that the proposed incinerator was "either completely trustworthy or a certain health threat," depending on which side a scientist was on, he said: "Being a layman is tough; you don't know if all these people know what they're talking about" (Michael Gibbons, 1986b).

Although TRASH continued its legal battles for another thirty months or so, critical decisions favoring the county over the township were made in 1987. The county had earlier asked the Pennsylvania Supreme Court to take jurisdiction in the incinerator dispute, and the same court subsequently ignored Plymouth Township's request in November 1986 that it refuse that invitation. Costs for the township's court battles were becoming a major factor by this time, because Plymouth had spent $120,000 in legal fees fighting the incinerator in 1986 and reluctantly hired a Philadelphia attorney at $250 an hour in 1987 for additional legal counsel in appealing the ruling ordering the township to cease all opposition to the county's incinerator plans. The state Supreme Court, in April 1987, ordered the Commonwealth Court to conduct "a prompt hearing on the litigation," and the latter court subsequently abbreviated the time permitted Plymouth Township to file its briefs (Margaret Gibbons, 1987). Between early June, when the Commonwealth Court heard the case, and September, when it upheld the lower court ruling permitting the county to construct the 1,200 tpd incinerator, the DER went ahead and issued construction permits in July for the facility. Despite additional petitions to the Commonwealth Court by Plymouth Township, appeals to the state Environmental Hearing Board by TRASH, and even the potentially disruptive decision by the Dravo Corporation to leave the waste-to-energy field in December 1987 (necessitating a change of vendor), county officials had every reason to be confident that an incinerator would be built in Plymouth.

The township and the TRASH protest group worked together, however, in legal attempts to thwart the county's plans throughout 1988, but the same judge handled all these cases and decided each in favor of the county. The June 1988 case was typical:

> Lancaster Senior Judge Wilson Bucher yesterday handed down a strongly worded order directing the township to approve the county's land-development plan and to "expeditiously" issue all the permits needed.
>
> The judge, specially appointed by the state Supreme Court to handle the litigation between the township and county, ruled that the county would not have to seek a variance or special exception from the township's Zoning Hearing Board.
>
> "The time has come, if it has not long since passed, for the township to cease its foot-dragging, nit-picking and logic-chopping," Bucher wrote in his 71-page opinion. "If the county is not granted prompt and comprehensive relief, its string of legal victories in this prolonged litigation will be indistinguishable from defeat." (Margaret Gibbons, 1988a)

Despite these initial legal victories, however, county officials said they would not give permission to begin construction until all of the litigation was resolved. Another problem that county officials did not emphasize in their interviews with the press was that the company originally scheduled to build the incinerator had been unsuccessfully attempting to sell its waste-to-energy subsidiaries. The latter difficulty was solved in principle during July 1988, when county officials announced that Montenay International Corporation of Miami, Florida, would build the incinerator (Margaret Gibbons, 1988b). In October, the state Supreme Court refused to hear Plymouth's arguments against lower court rulings prohibiting the township from attempting to regulate plant operations through zoning ordinances, prompting Bartle to comment: "This denial is a very important step in our three-year quest to locate a trash-to-energy plant in Plymouth Township" (Margaret Gibbons, 1988c).

The impact of recycling on the waste stream was never factored into the Montgomery County incinerator planning. Ignoring a November 1988 warning by a consultant, invited by Plymouth Township officials to speak to the assembled representatives of municipalities in the county, that new state laws "changed the entire ballgame for solid waste disposal" by mandating recycling of certain materials and thus dramatically reducing the expected waste stream (Weiss, 1988), Montgomery County signed an agreement with the Montenay Corporation in January 1989. Unlike Dravo, however, Montenay expressed an unwillingness to accept responsibility "for uncontrollable risks," contributing to an increase in costs from $137 million to an estimated $170 million—moving "tipping fees" from the originally estimated $36 per ton to between $60 and $75 per ton (Margaret Gibbons, 1988d). By April 1989, an Inter-Municipal Agreement (IMA) had been signed whereby other county municipalities had signed on to the project, and construction started in May 1989.

Costs for the project again escalated before it was completed in November 1991, after Plymouth Township and its TRASH activists had exhausted their legal options in attempting to prevent it. A November 24, 1991, account of the start-up reported, in part:

> The Montgomery County's trash-to-steam plant . . . passed its first test yesterday . . . when the county's fully computerized trash-to-steam plant began burning its first load of municipal waste. The $186 million facility . . . will test burn municipal waste for a few weeks, and plans to be in full swing, burning 1200 tons per day, by mid-December. . . .
>
> Twenty-four municipalities in Eastern Montgomery County will be delivering their trash to the plant. . . . If all goes according to plan, the facility will generate 32 megawatts of energy per day. Four megawatts will be used to power the plant and 28 will be sold to the Philadelphia Electric Co.
>
> "We're burning enough energy to power a 6,000-home community for an entire day," said plant manager Duane Wills. "As far as current technology goes, this is the best and cleanest way to get rid of trash."
>
> But the actual burning of trash at the facility, located on Alan Wood Road near Ridge Pike, has taken more than six years of legal and political wrangling. The plan has been beset by numerous lawsuits, two fatal accidents and loud outcries from environmental groups who claimed that the plant would emit hazardous levels of dioxins.

> The plant can burn just about anything, but Wills said "white items," such as refrigerators and washing machines, would be taken out. Also, aluminum cans and glass cannot be burned. Bits and pieces of these items are collected from the ash and taken for recycling, Wills said.
>
> "If the public is responsible about discarding their garbage, everything will work perfectly," Wills said. "We can only be as strong as the weakest link." (Banchero, 1991)

The Montgomery County activists were unwilling to complete separate questionnaires, claiming they feared individuals might be singled out for retribution by Bartle and other county elites supportive of the incinerator project. Some claimed this had already happened to "a few" opponents who had spoken out. One activist hosted an informative focus group discussion where opponents spoke about their own and TRASH's experiences over the years. No tape recorder or any notes were permitted at this meeting, lest some statement come back to haunt the speaker. One collective response to the activist questionnaire, which allegedly reflected the views of these Montgomery County opponents, was completed.[5] The collection of approximately 6,700 signatures against siting by the more than fifty activists who became involved politically was listed as one of the most important tasks accomplished by TRASH. Other types of work listed on the activists' questionnaire included "research on the issues," "speaking to school groups," "speaking to community groups," "testifying at PA House and Senate hearings," "writing letters," "speaking at meetings of other local municipalities," and "networking with other groups." In response to the items asking about "turning points in the group's development" and "decisions by the group or its leaders which turned out to be very important," TRASH's delayed decisions in taking legal action were emphasized. In response to an open-ended questionnaire item asking about "the most important consequences for you personally from involvement with the group," the follow-

5. Although at first refusing a request to complete questionnaires, a Montgomery County spokesperson agreed to complete a single questionnaire supposedly representing the opinions of the group after this focus group meeting upon learning that the random telephone survey, asking most of the same questions was already in progress. The reason for this change of heart seemed to be that activists might as well get their own perspectives on record to counter other opinions, but they would make it a collective response to protect their anonymity. Because such a questionnaire may be regarded as a single leader's perspective, at least, but may also provide valid insights of TRASH activists in some respects, it is used here with this caveat to the reader.

ing was written, presumably covering a number of individuals: "Burnout, stress, lack of personal time, disillusionment *especially with the judicial system*, disillusionment with DER, awareness of lethargy within community [emphasis in the original]."

During this December 1990 focus group discussion with TRASH activists, when construction of the incinerator had already begun, some reflected on the events of the past five years and what had been learned from them. One of the group's early mistakes, they acknowledged, had been focusing too exclusively on science and the experts while ignoring the major importance of politics in the whole dispute:

> We originally thought the plant could be beaten with science, the experts who were "carriers of the word," but we didn't realize until later that we really had to beat it politically because the courts had already been bought. Judge Bucher was assigned to all our cases, and he was in the industry's pocket.[6]

The Philadelphia Siting Attempt

Philadelphia's protracted siting struggle began in 1983 and continued until the summer of 1988, when City Council's Rules Committee voted the project down.[7] During those five years, thousands of local residents became involved in challenging the incinerator plans supported by Mayor Wilson Goode and other project proponents. The proposed site was at a United States Navy facility in the South Philadelphia area, which Jeff Horowitz, an attorney who worked in support of the project, described in an interview:

> I went down there myself, and the place where they were planning to build it was really a despicable little piece of property, tucked under a very high bridge off Route 95 on land the Navy didn't have

6. One condition TRASH activists established for this focus group discussion was that no quotations would be attributed to any individual.

7. The following individuals—including proponents and opponents of the proposed incinerator project in Philadelphia—provided useful information and insights for this analysis: Jennifer Nash, Janet Filante, Jim Walsh, Bruce Gledhill, Jeff Horowitz, Joe Cascerceri, David Cohen, Judy Cerrone, and Greg Shirm. Direct quotations from particular interviews are identified in the following paragraphs.

> much use for, and where nobody lived—more than a quarter mile from the nearest residents.[8]

The incinerator was to handle approximately 30 percent of the city's trash. Because this South Philadelphia community was not a separate political entity in the County of Philadelphia, however, no special compensation arrangements with local residents, in return for their accepting an incinerator, were legally possible.

Industry consultants commonly advise waste-to-energy site proponents to neutralize the predictable NIMBY opposition by getting as much initial support as possible from surrounding areas, and one of the suggested ways of accomplishing this is to establish a citizen advisory committee with significant input on siting and all related decisions. In Philadelphia, the site selection process did not include a CAC or any representatives of the residents living nearby. Rather, the mayor, a strong proponent of the project, met with vendors, lawyers, and consultants to decide on the location for the proposed incinerator, and only later would he appoint a panel of five Philadelphia physicians whose report favoring a siting would be dismissed by opponents as biased.

There was some early organized opposition to the incinerator project in Philadelphia in 1983, within months of its initial mention in the news. One of the first groups to oppose the plant was the Clean Air Council of Delaware Valley, which organized an early public hearing on January 28, 1984, with invited speakers—including Mayor Goode, an industry representative, and industry critics Karen Shapiro of the Center for the Biology of Natural Systems of Queens College and Neil Seldman of the Institute for Local Self-Reliance. Within two months, a loose coalition of South Philadelphia project opponents sent a letter to Mayor Goode, dated March 13, 1984, reading, in part:

> Before the City of Philadelphia proceeds any further with plans to build and operate the trash to steam plant, the City should:
>
> 1. Complete and publicize a survey of health conditions in South Philadelphia to determine whether community residents are suffering from excessive rates of cancer and other diseases which could be caused by toxic pollution, and
>
> 2. Determine the extent to which the plant would produce toxic

8. Interview with Attorney Jeff Horowitz, Philadelphia, January 18, 1990.

> emissions and the effect of these emissions on public health in South Philadelphia. . . .
>
> We would appreciate your response within ten days of your receipt of this letter.

It was signed by nineteen community leaders, with the contact person noted as Jennifer Nash of the Clean Air Council. A few months later, on June 20, 1984, some of the same spokespersons sent the mayor another letter along the same lines. The letter concluded: "We suggest that the City's solid waste disposal plans be made public now. A program to involve community leaders in the planning process should be developed. . . . We will contact you in the coming two weeks to discuss this matter."

One of the critical problems with this earliest challenge, according to activists involved, was that it was not a cooperative effort between black and white South Philadelphians. Greg Shirm of the Delaware Valley Toxics Coalition explained:

> The Clean Air Council, the Delaware Valley Toxics Coalition, and the Public Interest Law Center were instrumental in organizing what eventually became the critical opposition to the plant. When it was seriously proposed by Mayor Goode, we began to look at it to try to figure out whether it was a good or bad thing for the area. We began to "quick load" information from California and other places in about 1984 or 1985. . . . We received an early call from a Black Muslim group in South Philly which had questions about the incinerator, and they wanted to know what we knew about it. . . . So we held a meeting at the South Philly Community Center after sending out notices to community groups in South Philly. The turnout was mostly white, although some of the Black Muslims came. About thirty or forty people came to that first meeting we scheduled, and they were predominantly white. . . . But then for some reason the Black Muslims backed away from further meetings, and this initial opposition to the incinerator became almost entirely white, consisting of two dozen or so neighborhood associations from South Philly. . . . Our position at the time was that we were not opposed to incinerators, so we did not attempt to coordinate any opposition. . . . We were interested in seeing people get control over the kinds of technology used, its operation, and related issues. . . . We told them

> that it was not in their interest to allow themselves to be perceived as South Philly NIMBY groups.[9]

Because Mayor Goode and more than half the residents of the city were African American, an all-white opposition to his incinerator plans was hardly the ideal political strategy in Philadelphia.

Such early protest efforts were later built upon, in coordinating effective grassroots opposition, by Joseph Cascerceri, the staff director for the U.S. Congressman Thomas Foglietta for Philadelphia's First District. A full-time salaried activist who had moved into South Philadelphia in the mid-1970s to develop a community center, Cascerceri became involved in challenging the incinerator along with the neighborhood groups.

> I studied the [incinerator] proposal's effects upon our community, and came to realize it would be unhelpful physically, environmentally, and fiscally. I began to talk to community leaders from groups throughout South Philly about the issue, and we formed the Trash-to-Steam Alternative Coalition. I realized that we could not fight this as one section of the city against the rest, so we took the coalition and went to community meetings throughout the city in 1984, 1985, and 1986—two or three nights each week! . . . We studied alternative methods to mass burning and then made a presentation to members of City Council. . . . We invited them all to an auditorium in Center City one evening in the Fall of 1985, and eleven of sixteen attended. During this time we kept sending letters to the mayor and his staff, inviting them to these meetings or offering to bring information to his office, but he never responded.[10]

Clean Air Council member Jennifer Nash emphasized the difference between this new anti-incinerator coalition and her own group's less successful mobilizing efforts: "This neighborhood coalition of groups against the incinerator brought blacks and whites together, something the Clean Air Council had been trying unsuccessfully to do for quite a while prior to the incinerator project."[11]

An early indication of serious political problems for the project came in

9. Interview with Greg Shirm, Philadelphia, July 12, 1990.
10. Interview with Joe Cascerceri, Philadelphia, January 18, 1990.
11. Telephone interview with Jennifer Nash, January 12, 1990.

January 1985 when Philadelphia's City Council rejected the Goode administration's plan to build an incinerator:

> After nearly two hours of discussion and an attempt by Republican Council members to delay a decision, Council voted 15–2 to adopt a Council Rules Committee report that described the $151 million trash-to-steam plant as an "unnecessary risk to the health and safety of the residents of Philadelphia."
>
> Council's decision yesterday created a waste-disposal stalemate between the mayor and Council and makes it necessary for Mayor Goode to negotiate with Council. (Sutton et al., 1985)

Later that same month, Goode had resolved this issue by rejecting the proposed plant, although he also emphasized that he was just starting on a plan of his own, which would emphasize refuse derived fuel (RDF) plants (Loeb, 1985). Critics insisted that Goode's alternative would cost the city even more than the plant he had rejected, and the Philadelphia business community made its first public break with the mayor by insisting that "the best way to handle the city's huge volume of trash was to build a mass-burning trash-to-steam plant" and that "RDF just does not match mass-burn" (Cooke, 1985). By the end of 1985, the mayor's RDF plans had been dropped, and in early 1986 business leaders recommended reviving the old plan for building an incinerator at the Philadelphia Naval Shipyard. An editorial in the *Philadelphia Inquirer*, April 27, 1986, criticized Goode for his vacillation on the incinerator project—which the editors themselves supported—insisting that in the midst of the city's trash crisis mayoral leadership was necessary: "If Philadelphia doesn't get on with it soon, there won't be much health left to worry about—public, economic, or otherwise." In late April 1986, Mayor Goode announced that he was going back to his original plan to construct an incinerator near the Naval Shipyard in South Philadelphia (Kennedy, 1986).

Within a few days of that announcement, residents of South Philadelphia were again publicly protesting the proposed incinerator.

> From a Little League field at 19th and Bigler Streets to a touch football game in Packer Park, South Philadelphians reacted bitterly yesterday to Mayor Goode's renewed support for a mammoth trash-to-steam plant near the Naval Shipyard.
>
> Virtually to a person, residents of South Philadelphia said during

> interviews that they opposed the giant waste-burning facility when it was proposed a year ago, and that they had no reason to change their minds now. . . .
>
> The mayor's announcement Saturday also drew opposition from South Philadelphia neighborhood leaders and from five members of City Council. Four other Council members said they were undecided. One Council member, Republican W. Thacher Longstreth, said he supported the plan. . . .
>
> Studies have noted that South Philadelphia has the highest cancer rate in the city.
>
> The proposed incinerator, which would burn 630,000 tons of the city's residential trash a year, would produce steam to generate electricity that would be sold to the Naval Shipyard. When the mayor endorsed the trash-to-steam idea in 1984, reaction was so severe among City Council members and neighborhood groups that he changed his mind and decided to support a plan to build four small plants in different sections of the city. Those plants would have converted trash to fuel pellets.
>
> With that proposal at a standstill, the mayor went on to support a plan to construct a large trash-to-steam plant in Berks County—a plan that has become mired in legal difficulties and has faced mounting criticism from Berks County residents.
>
> Those events led Goode on Saturday to announce the revival of his first proposal—the trash-to-steam plant near the Navy Yard. (Eshleman, 1986)

Critics noted that Goode failed to mention recycling at all in his revival of his earlier proposal and called upon him to institute an immediate program to recycle "at least 25 percent of the city's solid waste by 1989 and higher levels thereafter" (Schirm et al., 1986). Within a couple of weeks, Goode's administration began orientation sessions for a few new workers responsible for starting the city's recycling campaign in some community-based projects.

On June 25, 1986, Mayor Goode made a televised appeal to city residents for support of the construction of a Navy Yard incinerator, insisting that trash disposal "has grown into a crisis that will only get worse unless we act now." Claiming that there was less than three years of total landfill capacity in the entire region, Goode criticized Councilman-at-Large David Cohen for opposing the plant—while Cohen became increasingly popular with South Philadelphia activists in the Trash to Steam Alternative Coalition because of his outspoken criticism of Goode's television pitch. Vince Termini was one of these activists who worked with Joe Cascerceri and other opponents in challenging Goode's project.

> "We're not going to trust any health expert the mayor hires," declared Termini, a bakery owner who is president of Girard Estates Area Residents (GEAR), one of the South Philadelphia civic associations whose strenuous opposition to the incinerator killed the plan in City Council two years ago.
>
> Judging from the reactions in the Termini household last night, . . . the mayor is headed for a major showdown with the community. Termini called the speech a "desperate act" and said that Goode didn't have the votes in Council to pass it. "If they do pass it, we'll take it to court and tie it up two, three, maybe four years," he said. . . .
>
> As to the panel of experts, Termini said he knows scientists who believe that the levels of dioxin the plant would emit would be a health hazard, increasing the risk of cancer. He said he talks on the phone now and then to Barry Commoner, the biologist who has written on the dangers of trash-to-steam incineration. "So why would we believe the mayor's experts?" Termini asked. "He's already decided he wants the plant." . . .
>
> A neighbor of the Terminis, Pat Nasuti, a nurse, said that in taking around petitions opposing the plant, even the most conservative, nonpolitical types had signed. (Cass, 1986)

The *Philadelphia Inquirer* officially supported Mayor Goode's incinerator proposal and criticized City Council for opposing it. This was the same newspaper that supported Montgomery County's efforts to build an incinerator in Plymouth in the previous case study. Its editorial on October 18, 1986, for example, referred to "the trash shell game that passes for Philadelphia's quest to find a safe, long-term, affordable way to get rid of its trash," complaining that the city was wasting millions annually on consultants and alternative trash disposal methods instead of going ahead and building the required incinerator. The same paper, however, was evenhanded in also criticizing an apparent attempt by project proponents to minimize critical public input. When the announcement of environmental impact hearings on the proposed plant was poorly publicized in November 1986, an editor of the

Philadelphia Inquirer complained along with outspoken opponents. "We had a serious concern as to why this was kept quiet until less than a week beforehand," Joe Cascerceri told a reporter. "They didn't give us enough notice to organize, so it will appear there's no one interested." An *Inquirer* editorial, November 4, 1986, was similarly critical:

> The brass at Philadelphia's Naval Shipyard has managed to confirm the worst fears of neighbors who think a trash-to-steam plant is going to be shoved down their throats. They did that by taking only the most legalistic, low-profile measures to publicize a public hearing on whether to sell the city a piece of Navy land for the project.
>
> For the record, that hearing is set for 7 p.m. Thursday at the Spectrum—a fact that only the most eagle-eyed citizen would have been able to spot in the Federal Register or buried in *The Inquirer*'s classified section.
>
> There are enough real concerns about the plant—its health and safety implications, for starters—without throwing a monkey wrench into the opinion-gathering process. As things stand now, it was left to long-time plant opponent, City Councilman David Cohen, to do what the Navy didn't: Alert South Philadelphia last week that the hearing was only days away.
>
> Mr. Cohen has succeeded for years in delaying Council's own hearings on the plant, . . . but his complaint about the Navy's silent treatment is no less valid.
>
> Officially, the aim of the Spectrum hearing is to assure "an early and open process" in shaping the scope of an environmental impact report on how a plant would affect traffic, air quality, energy conservation, and the welfare of the neighborhood—including the shipyard itself. And that's tough to do if no one shows up.
>
> *The Inquirer* has backed the concept of a trash-burning plant, if it can be operated safely and reliably. But it has backed, just as strongly, a process that guarantees real community input on questions of design, monitoring and adherence to pollution standards.

Despite such logistical challenges for grassroots organizers, however, within a few days they mobilized hundreds of angry residents to appear at the Spectrum hearing on November 6, 1986, to protest the proposed $280 million incinerator that the Goode administration was proposing to build on a 28-acre parcel of land along the Delaware River. It was the first major public meeting on the building of the plant, and the testimony received was to be used in an environmental study by the Navy in deciding whether to sell the land for the plant to the city. Bruce Gledhill was intimately involved with the Goode administration's project planning at the time, and without mentioning that the event was not well publicized he recalled later that they expected a massive turnout, prompting their rental of this major sports complex: "Negative voices always make the loudest noises, and we were deceived by them because we expected thousands but got dozens. We could have fit them into a high school auditorium."[12] In spite of the poor publicity, however, supporters of the project were greatly outnumbered by opponents in the audience.

> An angry and impassioned group of about 250 South Philadelphia residents and city politicians showed up at the Spectrum last night and bitterly denounced the proposed trash-to-steam plant at the Philadelphia Naval Shipyard.
>
> "I'm completely disheartened," said Judy Cerrone of South Philadelphia. "I feel helpless. I feel my son's health and welfare are at stake. As a mother, I feel like if it really comes down here, I'm going to get my son out of here. I don't want my son getting cancer."
>
> Spurred by the crowd's cheers, . . . state Sen. Vincent J. Fumo (D. Phila) declared, "We simply do not want this facility here." Fumo continued, "I can be a very formidable opponent when I want to be, and I want to be in this instance. I personally promise you we will tie you up in court at the state level, at the federal level, for years." . . .
>
> At one point, the irate crowd drowned out the words of Henry H. Reichner, executive vice president of the Greater Philadelphia Chamber of Commerce, who was one of the few speakers to support the proposed plant. "Where do you live? . . . Put it in your back yard," yelled members of the crowd as Reichner spoke. (Montaigne, 1986)

Throughout the Philadelphia challenge, national anti-incineration spokespersons were also instrumental in persuading some municipal officials to join in actively opposing the trash project supported by the mayor. Barry Commoner, Paul Connett, and representatives from Greenpeace as well as from the Environmental Defense Fund were among the outsiders

12. Interview with Bruce Gledhill, West Chester, Pennsylvania, January 19, 1990.

who played roles in helping legitimize opponents' challenge among Philadelphia politicians and the general public.

Despite the increasing opposition, the Goode administration continued its efforts throughout 1987 to site the trash plant. Some incinerator advocates claimed that their polls were showing support for the project in many parts of the city, and surprisingly only a 50–50 split even in the South Philadelphia area closest to the plant.[13] Opponents offered an assortment of more-or-less serious alternatives ranging from mandatory recycling to burning it in nearby Chester to shipping the city's trash to Curaçao. An editorial, for example, in the *Philadelphia Inquirer*, January 2, 1987, mocked Councilman Cohen and City Council for vacillating from alternative to alternative in finding a way to dispose of Philadelphia's trash without building the incinerator:

> Chester's offer to take Philly trash has been looking shaky lately. But not to worry. That was 1986's gambit. For 1987, Councilman David Cohen and chums have pulled a new one out of the hat: Trash-to-Curacao. Yep, you heard right. A shipping line has inquired about hauling city trash to Curacao, an island off Venezuela where . . . it would be burned to generate power.
>
> Trash-to-Curacao is Mr. Cohen's latest flirtation. It follows trash-to-coal-mines (rejected by upstaters), trash-to-Ohio (frowned on by Ohioans), trash-to-Berks County (vetoed by Berksers), trash-to-confetti fuel (rejected by the marketplace), and trash-to-Chester (hung up by siting hassles).

The City Council Rules Committee dealt a major blow to Mayor Goode's incinerator proposal in late January 1987 when it refused to send a bill to the full Council for a vote after a draining day of analysis, debate, cheers, jeers, and misunderstandings:

13. Such claims were made in the course of interviews with former Deputy Commissioner of Streets in Philadelphia Bruce Gledhill (January 19, 1990), and Jeff Horowitz, an attorney for the proposed vendor, Ogden-Martin (January 18, 1990). The local researchers reporting these results, however, were unwilling to share copies of their survey instruments or any of the relevant data. A comparison with our backyard data suggested either that such claims were mistaken or that the percentages had changed considerably in the neighborhoods near the proposed site by the time of our survey. Only 18 percent of the Philadelphia backyarders we contacted by telephone said they favored the siting of the incinerator.

> The committee voted unanimously to table the bill until the Goode administration considers an alternative plant being developed by officials in Chester. The committee had been widely expected to clear the bill for a Council vote that Goode has predicted would bring approval of the $280 million project he had touted as the solution to the city's trash crisis.
>
> "The project is dead," a jubilant Joseph Cascerceri, leader of the opposition movement, told reporters minutes after the vote. "I'll predict that right now. There is no way that trash-to-steam plant is ever going to South Philadelphia after today."
>
> Councilman Edward A. Schwartz, a member of the committee and a supporter of trash-to-steam, said the committee's vote did not necessarily signal the demise of the project. . . .
>
> The committee action came after Councilman Lucien E. Blackwell accused Deputy Mayor John P. Flaherty Jr. of showing "disrespect" to Chester Mayor Willie Mae Leake by refusing to take seriously her city's proposal to spend $335 million developing a trash plant on the Delaware River.
>
> "If we cannot agree with her, that's fine," Blackwell said angrily, "but I will not sit here for one minute and have you disrespect the mayor of the city of Chester. I will not do it."
>
> Blackwell's outburst drew cheers from several hundred trash-to-steam opponents who packed the gallery, many of them carrying protest signs in the shape of tombstones. . . .
>
> Flaherty, who was obviously disappointed by the turn of events, said later that he had meant no disrespect to Leake. He said he would discuss with Goode the committee's request to explore the Chester alternative. . . .
>
> In a separate interview, Blackwell said Flaherty "blew it" by being "very arrogant and very nasty in dealing with people regarding trash-to-steam. They're not lord and master of what goes on here," he said, referring to administration officials. "We're the legislative body, and ultimately we'll make the decision. . . ."
>
> Leake urged Council members to "respond to a neighboring city in distress" by sending all of Philadelphia's trash to the 6,000 ton-per-day plant it hopes to develop. The Philadelphia plan calls for burning 2,250 tons daily. (Clark, 1987)

In his testimony, Flaherty had referred to Leake's plan for a Chester incinerator—the preferred alternative of some municipal officials to the Delaware County plant discussed in Chapter 4—as a "mirage" because of serious questions about its timetable, disposal fees, and the problems Chester was having in even acquiring the land upon which it proposed to build its plant.[14] Yet the Philadelphia project continued to be related to outside projects such as Chester, because developers and others realized that Philadelphia's difficulties in siting a facility within its own boundaries could justify a larger facility than might otherwise be warranted outside those boundaries—to handle Philadelphia's trash, of course.

One of those intimately involved in discussions about the attempted siting in Philadelphia was Bruce Gledhill, the former Goode administration official whom Jeff Horowitz, attorney for vendor Ogden-Martin, referred to as "a lightning rod for the city." Well versed in the links between the Philadelphia and both Chester projects, Gledhill explained their relationships to Philadelphia's Navy Yard plant from the perspective of a former insider in the Goode administration:

> As Philly sought to find a solution for its trash disposal problems, it preferred to look inside the city. . . . But as it became apparent that there would be difficulties involved in siting such a large facility within the city itself, developers and others outside Philadelphia viewed it as an opportunity to justify a large facility which could not otherwise be considered. . . .
>
> The Philadelphia and Chester projects were linked because there was a need to find a disposal solution for Philadelphia and a recognition by others that it was going to be difficult to site something within Philadelphia. . . . Interest in out-of-county solutions waxed and waned with Philadelphia's own preferred alternatives. . . . David Cohen and Philadelphia's City Council kept outsiders informed and worked with political entities outside the city to find an alternative solution to Philly's trash problems. . . .
>
> There were two Chester projects, and they approached Philly in different ways. The Chester city project . . . has a long and torturous history. . . . There is now a federal investigation over the whole thing: it became a project in search of trash and not trash in search of a project—merely a way of bailing out a bankrupt community which is among the most desperate in Pennsylvania or in the whole United

14. In fact, there were so many financial and legal problems with this city of Chester project that it was eventually dropped.

> States, for that matter. . . . So, City Council in Philadelphia was sympathetic to Chester's plight while it was also looking for a disposal system. . . .
>
> How much did Cohen's telling outsiders that Philly's project was dead drive a search for some outside facility? This comes under the category "hearsay," because I wasn't directly involved in such communications, but the links between Chester's development team and Philly's City Council were very strong. . . . Even in 1984 and 1985, whenever there was a meeting on the incinerator involving Philadelphia City Council members, there would also be people from the Chester development team present, and at the same time there was a consultant working for City Council, Neil Seldman of the Institute for Local Self-Reliance, who was also working for the city of Chester. . . . There had to be frequent communication—otherwise, there was no way the city of Chester would have considered such a large incinerator, because the city itself generated only 100 tons per day. They had to be looking from the outset to Big Brother Philadelphia to come to their aid in supplying the trash.[15]

Observers as well as politicians on both sides of the issue viewed late winter and spring of 1987 as "critical" because of the upcoming May 1987 primary elections, when the mayor and all the Council members were seeking their party's renomination, and also because the budget had to be passed by the end of May. Philadelphia's Mayor Goode put in what appeared to be merely a cosmetic visit to Chester's Mayor Leake in early 1987—at the urging of Philadelphia's City Council members, who were disturbed by the alleged insult of the Chester mayor mentioned above. Leake reported that Goode told her "it might be possible that [Philadelphia] would send some of its trash to her Delaware County city" but also insisted on building his own plant at the Navy Yard (Sutton, 1987). Voted out of City Council's Rules Committee in February 1987, the Philadelphia incinerator project became the focus of heated public hearings, debates, and protests for the following eighteen months.

Opponents stepped up their efforts throughout 1987 and into 1988, busing thousands of people to City Hall with signs, gas masks, and other props

15. Gledhill interview, January 19, 1990, when Gledhill was working for Weston after leaving his job as Deputy Commissioner of Streets in Philadelphia in December 1989.

for each of these public hearings. Joe Cascerceri summarized these mobilization techniques for us:

> There were eleven public meetings, and at each we took residents to chambers of City Council—sometimes as many as 3,000 people. We packed those chambers with signs, gas masks, and surgical masks—all kinds of gimmicks to make our case. We received good media attention, and good exposure at City Council.
>
> In the months before these meetings, we assigned twelve neighborhood leaders, all volunteers, to go on a weekly basis and make appointments at the offices of the various Council members and sit with them to lobby for our position through slides, brochures, and handouts about alternatives to the mass burn facility. . . . We had phone banks in the community centers of each group. . . . We went each week during the winter to senior citizen meetings with paper, envelopes, stamps, and pens, encouraging people to write letters to City Council and the mayor. . . . We asked each person to pick three of the twelve reasons we listed to oppose it, so they wouldn't all be the same. . . . During April and May of 1987, City Council was flooded with over 30,000 letters. . . .
>
> At that point, City Council delayed a vote and went back to public hearings which, unfortunately, meant more work for us. . . . So during the summer and fall of '87 we had to organize the people, get buses, solicit donations from businesses, and all the rest. . . . Termini's Bakery—owned by Vince Termini, one of our coalition's members and a very active civic organizer in South Philadelphia, for example—supplied free lunches to everybody, costing $4,000–$5,000 over a five-month period.

In addition to such grassroots organizing, the opposition also had the backing of Councilman Cohen, whom Cascerceri and others characterized as "a champion of the rowhouse person in this city." Cohen's support in City Council was critical for the opposition forces. Julius Curry, a black activist who helped mobilize some neighborhoods against the project, emphasized Cohen's influence with certain black members of City Council in focusing attention on environmental issues and alternative technologies rather than allowing it to be framed as black mayor versus white NIMBYs.[16]

16. Julius Curry was present at the focus group discussion in the home of a Philadelphia activist on July 9, 1990.

Cascerceri's protest coalition had strong connections with Cohen, a politician for whose reelection they had worked hard.[17] Although Mayor Goode did not support Cohen, who ran on the same Democratic ticket as he in his bid for reelection, Cohen won by the largest margin in the city's history. Cohen emphasized his own inclination to trust "people in the community over the so-called experts," explaining that his bias derived from years of experience:

> You've got to work with communities if you are their representative. And even if you could demonstrate factually that a community's beliefs are dealing with perceptions rather than reality, those perceptions which are deeply held are themselves a reality. . . . And so if people live in terror of emissions from a plant, that plant will destroy their quality of life—even if it were later to be proven that nothing went into the air from the plant! . . .
>
> Most people feel that you can't fight government, because it's so powerful. In the beginning of the fight in Philadelphia, I met with the mayor simply as one of the members of the community groups. Even though I was the councilman, they asked me to join them.[18]

Discussing the area's coercive and undemocratic political machine, which had accustomed residents to authoritarian politics, Cohen observed:

> My main hurdle was first to convince the people of South Philly that they could win this fight against trash-to-steam. That was my main hurdle because the people, particularly in South Philadelphia, were so convinced that it could not be won. . . . You can never guarantee outcomes, but it was my strong conviction that we could win this one if we stayed united, because we were right. . . . There's so much internal warfare in both political parties that the people often have the feeling that the politicians start out by being on your side but then, when pressure is applied to them or other inducements are found, they come back to you and say, "Well, we've got to support it."
>
> My entry into the case apparently created a difference, because my

17. In the middle of his own interview with Ed Walsh, Joe Cascerceri used his telephone to call Councilman Cohen's office and schedule an interview the following day for Walsh.

18. Interview with David Cohen, Philadelphia, January 19, 1990.

> reputation is well-known, even though I never until this last election got any support out of South Philadelphia 'cause I was a rebel, a maverick, a liberal. I had too good relations with blacks, or I was always fighting for minorities, and therefore I was fighting against them. That's how I was viewed. . . .
>
> As time went on, I became a kind of a center and all of the political forces came together around me. I sort of helped create a situation where what usually occurs—you know, politicians fighting with one another and then the whole thing collapsing—didn't occur, but we kept getting stronger. Well-established politicians in South Philadelphia have told me, "You know, we could never have won this fight without you, because we'd have all ended up fighting with one another and the mayor would either go through us or get all of us together on his side, but you made that impossible to occur." . . .
>
> South Philadelphia had a dinner for me. . . . I was the only person the community felt was "absolutely and forever." Now, I'm not saying that by way of criticizing others. Maybe they were more practical than me. People were warning me that this is going to cause you to lose your Council seat because the liberal progressive community, with which I was associated, couldn't understand why I didn't see the progressive character of incineration.

Cohen's importance in the struggle was widely recognized among both opponents and proponents. One of his own favorite stories was about the *Philadelphia Inquirer* columnist who wrote, after the project was defeated, that he never thought he'd live to see the day when a Jewish politician like David Cohen would rival former Mayor Frank Rizzo as the most popular public figure in heavily Italian South Philadelphia.

Both Councilman Cohen and Mayor Goode were reelected to new terms in the November 1987 elections—the mayor by a slim margin of victory and Cohen by an overwhelming margin. Goode, however, renewed his push for the incinerator, resubmitting his plan to City Council, where it had twice previously faltered—again precipitating strong opposition. Widespread political pressures from citizen groups around the city, which Joe Cascerceri called "the largest grassroots opposition Philadelphia has ever seen," coupled with Councilman Cohen's leverage and resulted in a key victory for the opponents, because the incinerator proposal never got to the floor for a vote. Events at a mid-April 1988 hearing were typical:

> For a few raucous hours Wednesday, City Council's chambers became a sort of convention hall, with neighborhood groups from across the city packing the floor and balconies to protest any move to put trash plants near their homes.
>
> The occasion was a Council hearing to consider possible solutions to the city's trash crisis. As expected, the featured attendants were the people of South Philadelphia who urged Council to reject Mayor Goode's perennial proposal for a trash-to-steam plant at the Navy Yard. (Miller, 1988)

In the course of this meeting, Cascerceri was cheered wildly by the audience when he vowed in his statement to stop the incinerator project by any means necessary: "Enough is enough," he said, "we cannot and will not allow a trash-to-steam facility to be built at the Navy Yard." And Northeast Philadelphia residents came in busloads to protest City Council's plan to investigate other possible sites for trash-to-steam, including two sites in their part of town, along State Road (Clark, 1988).

In July 1988, Mayor Goode dropped the plan for lack of support, precipitating victory celebrations by opponents. An editorial in the pro-incinerator *Philadelphia Inquirer* for July 29, 1988, summarized events under the head "Demise of the trash plant requires city to gamble":

> So this is the way the trash-to-steam plant crisis ends. Not with a bang, but with an exchange of letters between Mayor Goode's office on the second floor of City Hall and City Council President Joseph E. Coleman's office up on the fourth.
>
> The mayor wrote Mr. Coleman on Monday saying that the project had "become inoperative through lack of Councilmanic leadership." Mr. Coleman replied promptly . . . , confirming that the plan has indeed become "inoperative," and stating that Mr. Goode's letter "is an attack . . . upon the leadership I provide to the Council."
>
> Council's refusal to vote either for or against the trash-to-steam plant, . . . which had seemed to be the most effective and logical plan available, has to be seen for what it is—a high stakes gamble. . . .
>
> To be sure, Alices are reporting from various Wonderlands to suggest that Philadelphia can solve its trash problem by developing a recycling program that would be far more effective than that of any

> other city in the country, or that an escape is to be found in the City of Chester's still-unbuilt trash plant.

Additional insights into the complexities and the internal dynamics of this Philadelphia grassroots struggle came from responses to our activist questionnaires and focus group discussions. We also encountered the solidarity among themselves and distrust of outsiders that bound these activists together and helped them achieve their goals. Seven had already completed our questionnaires before one woman refused and aggressively discouraged others, arguing, despite our assurances to the contrary, that the information might be used to make another siting attempt in the area. The same person, however, invited us to her home for a focus group discussion with most of the people she had convinced to refuse, provided we agree not to tape our conversations. Responses on the questionnaires revealed the diversity of the opposition base, because in response to a question asking "the name of your group," for example, four different groups were listed. The "types of work usually done for the group" covered the wide variety of activities we would expect from these successful grassroots activists: "meeting city officials," "organizing rallies," "lobbying," "demonstrating," "making phone calls," "painting posters for protesters to carry in City Council," "hosting neighborhood meetings," and related political activities. Somewhat overlapping responses came from questionnaire items asking about "turning points in your group's development" as well as "decisions by the group or its leaders which turned out to be very important." Key turning points emphasized were the early decision "to form a broad-based citywide coalition" and the obviously related political opportunity created by it: "City Council's opposition to the Mayor." They also mentioned "intense educational efforts with civic groups throughout the city" and "lobbying efforts with government officials."

These Philadelphia activists were a much larger and less homogeneous group than their Montgomery County counterparts, and it was virtually impossible to bring them all together for a discussion of their experiences. In the process of arranging one focus group discussion during July 1990, we were given insights on the complex nature of this coalition of grassroots groups. Referring to the key persons in the citywide struggle as "the 12," neighborhood activists included five of the seven from whom we had already received questionnaires. Yet a number of their "12" were not included on the lists of "key activists" by their more cosmopolitan colleagues, from whom we also collected names. Such discrepancies are hardly surprising for such a broad based coalition of activists (see Walsh and Warland, 1983).[19]

One non-neighborhood activist emphasized the evolution in the public perception of neighborhood organizations:

> It's certainly the case that the South Philly groups were made to seem as if they were NIMBYs who simply didn't want the incinerator in their area. What I saw over time was an evolution in their thinking about how they should approach this. . . . Whether that change was sincere or not, I have no way of knowing. . . . Later in the struggle, around 1987, they realized the situation they were in if they simply said "We don't want this thing in South Philly," so they began to look for alternatives.[20]

All the activists agreed that Councilman Cohen's contribution was very important in persuading progressive black members of City Council to oppose the incinerator, without whose support the predominantly white protest movement could probably not have successfully challenged this pet project of Mayor Goode.

While such insights relativized the importance of the neighborhood protest groups, none of the non-neighborhood activists attempted to minimize the critical role of such groups. Cohen, in fact, worked closely with them. Key black neighborhood leader Julius Curry explained one of his own mobilization strategies in the focus group discussion. His approach was to agitate incinerator proponents into making statements that could then be used selectively with different audiences to foster a public perception of duplicity or worse.

Overview and Continuities

The Plymouth siting in Montgomery County was obviously facilitated by its location in politically centralized Montgomery County and initially sup-

19. Two of the activists whose names were given to us by Joe Cascerceri, and from whom we received questionnaire responses, were not included among "the 12" by the activists attending this focus group meeting. A number of people on "the 12" list were not considered central activists by Cascerceri and others from whom we accepted names. Most noteworthy is that the woman who encouraged some others not to complete our questionnaire was not included on the original list of activists but was added when mentioned by one of those on the original list.

20. Shirm interview.

ported by officials in the backyard community attracted by the project's $350,000+ compensation package. Whether a quicker challenge by the TRASH activists might have prevented local authorities from agreeing to host the project, and whether such a refusal by local authorities might have been enough to dissuade Commissioner Bartle and other project proponents, we cannot know. Whether a broader-based challenge including numerous municipalities besides Plymouth would have been more effective is another moot question. South Philadelphia, on the other hand, was not a separate political entity from the rest of that city, and thus no special compensation package could be created to persuade local officials and residents to accept that incinerator. Grassroots opponents also more quickly mobilized a much broader base of support there. Different strategies and tactics were used by opponents at each site, with Montgomery County activists relying primarily upon experts and legal strategies rather than upon the political organizing favored by their South Philadelphia counterparts. The Philadelphia siting might very well have been successful in the absence of the full-time organizer and the councilman who promoted opposition both within City Council and outside the target area.

This is the first of our comparisons where various physical and structural characteristics of the project coincided even roughly with predictions. Specifically, the sited Montgomery County project was smaller than its defeated Philadelphia counterpart, was situated in an area of somewhat lower population density, and promised the host community significant compensation. Another favorable characteristic of this Montgomery County target area, from proponents' perspectives, was that it had no previous history of organized protest. The only variable of those listed in Table 6 on which Montgomery County seemed a less promising site than Philadelphia was trash importation.

Again, our fieldwork data for these two cases are useful in better understanding the survey results reported in Chapter 3. Despite data suggesting that Montgomery County backyard residents had higher socioeconomic status and more dense preexisting networks than their Philadelphia counterparts, the data also revealed that the Philadelphia backyarders had more negative attitudes toward incineration, which can be attributed to the effectiveness of its widespread grassroots political mobilization around the incineration issue. The Philadelphia anti-incineration effort, in other words, relied more upon creating neighborhood solidarity around this issue than on activating existing network ties or tapping into residents' existing resources.

Can we identify any emerging theoretical patterns among the three sitings thus far? Two site characteristics are similar across the Delaware, York, and Montgomery projects—one in the predicted direction, and one opposite predictions. In none of these three areas had there been previously organized protest that might have served as a source of activists as well as mobilizing skills. The more surprising commonality among the three, however, is that they all planned to import outside wastes—something observers consider threatening for such projects. In addition, no organized opposition from outside these respective backyard communities emerged, at least partly because of local officials' early support for the county project (Montgomery and York) or for incineration in general (Delaware).

And what about common patterns at the three defeated projects? In terms of the site characteristics, no citizen advisory committee was associated with any of these defeated ones. Opponents of these proposed sites also solicited and accepted assistance from groups outside the backyard areas in the Philadelphia, Lackawanna, and Cape May struggles, prompting each to frame its protest in countywide rather than more local terms. There were also splits among political elites associated with each of these defeated projects.

Do these two additional cases offer evidence on the timing issue raised at the end of the previous chapter about the possibility that the broader regional and national anti-incineration movement influenced the outcomes of local protests? Not obviously, because the defeated Philadelphia proposal became public in 1983, a year or so before the Montgomery County project (1984), and the first protests also came approximately a year after these dates in each case.

7

Two Last-Minute Legislative Defeats: Broome and St. Lawrence

Our original intent in selecting two New York state projects was to contrast a siting with a defeat as in the preceding chapters. We intentionally took two projects still in process. The one in Broome County appeared, even to local opponents, likely to be successful; the one in St. Lawrence County had suffered a major setback in the county legislature and seemed doomed. Although things turned out differently from what was expected, these two cases where citizen opposition was effective for different reasons tell us a good deal about the factors that promote and militate against such sitings.

The final votes in both of these cases came in the early 1990s, and we follow our established pattern by discussing the likely siting (Broome) first, even though its defeat came after the St. Lawrence decision. The startling Broome County vote took place in February 1992; the similarly surprising St. Lawrence vote took place nineteen months earlier, in July 1990.

Site Characteristics

Both planned sitings called for relatively small incinerators (see Table 7), with Broome's projected to be twice as large as that in St. Lawrence County.

Table 7 Characteristics of proposed incinerator sites in Broome County (N.Y.) and St. Lawrence County (N.Y.)

	Broome Co.	St. Lawrence Co.
Size (tons per day)	571	250
Out-of-county wastes to be accepted?	Maybe	Maybe
Estimated population within 2-mile radius	2,000	12,000
Amount of host community annual compensation	$250,000+	$75,000
Citizen advisory committee for incinerator?	Yes	Yes
Any previous organized protest in area?	No	No

In both cases also, opponents claimed that out-of-county wastes would be required by the project, while proponents were noncommittal on this issue. Population density around the Broome plant, however, was considerably lower, while proposed host community compensation was also significantly higher. Both projects had what some informants referred to as citizen advisory committees associated with them, and neither had experienced any previous organized protest. Based only on the characteristics in Table 7, an informed observer might expect both projects to have involved successful sitings—with the acceptance of outside wastes being their only major problematic aspects. Such data do little to explain why both were defeated.

The Broome County Siting Attempt

The New York State Legislature created the Broome County Resource Recovery Agency (BCRRA) in 1983 as the implementing organization for siting a single mass-burn facility in the county.[1] The solid waste education coordinator in Broome County, Kathleen Hennessey, explained in a 1989 interview that recycling was not an important consideration at the time the incinerator was first being considered there:

> Under the former Republican administration of Carl Young, who took office in 1980, incineration was pursued as an alternative to

1. The following individuals, again from both sides in this conflict, provided useful information and interpretation in constructing the account of the Broome County siting process: Lori Dietz of the Resource Recovery Agency, Kathleen Hennessey, Lee Cornell, Audrey Glover, Charles Mead, Kathy Moreland, Chris Burger, and John Smigelski. Specific interviews or written communications used in the text will identify individuals and dates.

> landfilling. . . . It was to the credit of the Young administration that it was looking ahead to the time when the county landfill would no longer be able to receive trash. The technology which was being touted at the time was waste-to-energy, and there were really very few people who were talking about recycling as a component of solid waste management. . . . [Recycling] just had not been conceptualized in the early 1980s as a part of such a plan. It was still viewed as a voluntary activity for fund-raisers or fringe elements of the population. . . . So this incineration facility was planned with only a nominal amount of recycling considered.[2]

A consulting firm identified four possible sites for the Broome County legislature, and public meetings began. Charles Mead, a leading opponent at an original site that was later dropped, explained that he and his wife, in cooperation with others, used outside speakers and whatever literature they could find to "turn our town board around" in the Broome County municipality of Maine. Their local opposition group called themselves TOMORROW (Town of Maine Opposed to Resource Recovery of Waste).

> We started our fight in 1985 or 1986, when the county executive was in favor of the incinerator. Proponents of these things always take the path of least resistance. Our town board had originally agreed to have it here because they were looking at the kickbacks. . . . Only when we went out and collected 1,600 names did they finally turn it down.[3]

After similar early political opposition emerged at the other sites, one site in the Kirkwood area was selected by the county legislature in June 1986 as the preferred location for the incinerator. At least superficially, this process reminds us of the process in York County, where the Black Bridge site was selected only after all those initially listed refused the project. Although the process through which the specific decisions in this case were made is not public knowledge, the minutes of the town board meeting for as early as June 5, 1984, show that Kirkwood's council members voted affirmatively to "retain Robert J. Kafin, Esq. . . . of Glen Falls, New York, as special counsel to the town regarding the proposed Broome County Resource Re-

2. Interview with Kathleen Hennessey, Binghamton, New York, November 1, 1989.
3. Telephone interview with Charles Mead, Maine, New York, November 8, 1990.

covery facility." Some Kirkwood residents claim they were deceived by town officials, who assured them from the beginning they would resist any siting, and these skeptics also suspect that financial considerations overrode any initial intention Kirkwood's civil leaders may have had to resist the project.[4]

Debate and discussion between county and Kirkwood officials continued from 1986 to 1988. During 1986, Requests for Proposals were issued, a draft Environmental Impact Statement by the BCRRA was publicly discussed, and work was started on a health-risk assessment (HRA) related to the proposed site. Although both BCRRA officials and opponents of the Broome incinerator said there was a citizen advisory committee formed in conjunction with the project, the list of names from the BCRRA was different from those on the list received from Audrey Glover, a Kirkwood opposition activist appointed by her town's board as co-chair. Only three names were common to both lists, and further inquiry revealed the existence of at least two committees loosely referred to as citizen advisory committees, neither of which was very influential on the project. Commenting on the committee with which she was involved, Glover explained:

> We began raising questions about the proposed plant, and asked the [Kirkwood Town] Board if they were against the project. We never got an answer, and then a few members dropped off because they felt a deal was being cut. . . . I, as co-chair, and a few other members began digging deeper as negotiations were coming to a climax. . . . It was in February of 1987 that my co-chair informed me that the supervisor told him to "keep a lid on the committee so they don't interfere with the negotiations." . . . Kirkwood officials then began saying that they had no choice, that it was a county decision, and that they were just going to get the best deal possible for Kirkwood. . . . The next thing we knew, we were abruptly dissolved, and until this day there are no board minutes depicting this as a resolution. . . . Our committee was only in existence for eight months, from September 1986 to April 1987. The CAC was just a facade.[5]

4. Broome activists and citizen opponents at other sites sometimes referred to illegal kickbacks they suspected public officials might be getting, but more often they discussed legal economic incentives these officials would be offered to favor these projects. Such major building projects in any county gave the officials in charge of them leverage over millions of dollars in contracts and subcontracts, for which they could expect to be repaid by recipients.

5. Audrey Glover, letter to Ed Walsh in response to an early draft of this chapter, November 6, 1993.

Because CACs are commonly viewed as organizations created by proponents to facilitate siting processes, we sent questionnaires only to the people on the list received from the Resource Recovery Agency's Lori Dietz. Responses were received from six of the nine people, including two of the three individuals who were on both Dietz's and Glover's lists. The following comments, received on our item asking "the most important accomplishments of the CAC," do not suggest that this committee played a major role in the process: "No important accomplishments," "Reviewed details of the health-risk assessment for the incinerator and tried to ensure that all reasonable inputs and potential health hazards were included," "Lobbied for the dissolution of the proposed trash burning plant which did not take into consideration recycling and other means of reducing trash," "Established emission controls and monitoring programs."

In March 1987, Broome County issued its final requests for proposals to design, build, and operate the incinerator. In the same year, the county also hired a recycling coordinator to head the first community recycling project in New York State and announced details of what it referred to as an "integrated solid waste management plan." Various recycling initiatives in different county municipalities were undertaken.

The New York Public Interest Research Group (NYPIRG) insisted that recycling should be given priority over other methods of disposing of trash and helped organize local resistance to the Kirkwood project in early May 1987.

> About 100 persons rallied yesterday near the site of Broome County's proposed garbage incinerator along Stratmill Road in the Town of Kirkwood. The rally, organized by the newly-formed Broome Citizens Against Incineration and NYPIRG, was held to publicize the groups' call for a five-year moratorium on construction of new garbage-burning plants and creation of a statewide Waste Recycling Authority.
>
> Broome County plans to build an incinerator in Kirkwood by 1992 that would burn 500 tons of trash daily. . . . Protesters carried signs yesterday reading "Don't Make Us Guinea Pigs," and "Incinerate [County Executive] Carl Young." They released 100 black balloons to symbolize dioxin emissions that allegedly would come from the plant. . . . The launching of the balloons is the start of a campaign to fight incineration. . . .
>
> At the state level, NYPIRG is urging a moratorium on construc-

> tion while the state develops a plan of mandatory recycling that would recycle 60 percent of the state's garbage in 10 years. . . . NYPIRG officials said the construction of expensive incinerators when successful recycling programs exist worldwide is poor public policy. . . .
>
> The new Broome County coalition will concentrate on a letter writing campaign to state lawmakers, [Joan Harris, project leader for NYPIRG] said. "Members will also attend hearings on the plant's environmental impact statement due to begin this summer," she said. (Basler, 1987)

A group of local Kirkwood residents began gathering information about approaches to solid waste management in May 1987, and they defined their purpose as an attempt to raise questions in the public's mind about the county's proposed resource recovery project via public meetings and newspaper articles. Despite the spring 1987 rally, local questioning of the project, and the outside urging of NYPIRG, it was not until December 1987 that a Kirkwood-based protest group was formally organized.

By July 1988, the Foster Wheeler proposal had been selected by the BCRRA from among six received, and this company was designated as the vendor to design and construct the incinerator. Opinions varied regarding the final agreement between county and local officials:

> A deal to make the Town of Kirkwood feel better about being the host community for a 500-ton-a-day garbage-burning plant could bring more than $6 million into town coffers.
>
> An undated draft agreement between the town and Broome County requires roughly $6 million in cash payments to Kirkwood over 25 years to offset potential property tax revenue that could be lost at the 96-acre site of the proposed incinerator.
>
> The town would also receive. . . . almost $45,000 to train and equip fire-fighters. . . . The county also plans to spend another $750,000 installing a new sewer and water district around the plant. . . . and pay up to $500,000 more should the town undertake improvements or expansion in the utility district. . . .
>
> The numbers are in an undated copy of a draft agreement obtained by the *Press and Sun-Bulletin*. The document was marked "confidential." County and town officials yesterday declined to comment

> on the draft of the agreement, which they have been negotiating for about two years. . . .
>
> Construction of the incinerator, expected to cost about $65 million, excluding financing and other costs, is scheduled to begin next year. (Sbarra, 1988a)

In August 1988, Broome County began legal action against nine Kirkwood property owners to take control of approximately sixty acres needed to build the incinerator.

> The county turned to the court after all nine property owners rejected buyout offers. Municipalities are empowered to take private property for public purposes under state eminent domain laws. . . . None of the property sought by the county has homes or other structures. . . . Owners interviewed last week said they rejected the county's offers because they were "ridiculously low." The county offered $600 to $1,000 an acre, according to a member of one family involved in the matter. (Sbarra, 1988b)

The BCRRA went on to sign a service agreement with Foster Wheeler in October 1988, and a host community agreement with Kirkwood was also finalized at this time. Permit applications for the facility were submitted to the New York State Department of Environmental Conservation in December 1988.

In their frustration over not being informed by the town board about its incinerator agreement, some Kirkwood residents formed the nucleus of the local opposition group that would eventually call itself RECYCLE (Residents Economically Concerned for Your County's Living Environment). This social movement organization worked loosely, at first, with another opponent group, Citizen Action of New York, to file a lawsuit against Broome County, charging that the county did not develop "a comprehensive solid waste management plan as mandated by law":

> RECYCLE has complained that the county failed to do a thorough environmental impact statement on its solid waste plan. The county contends its environmental review is thorough and that its solid waste plan is comprehensive. . . .
>
> County Executive Carl S. Young [touted the plan] as the long term solution to local trash disposal problems. . . . "Obstructionist groups

often appear on the horizon in this kind of project or any major project," he said. He doesn't think the groups will delay the recycling or resource recovery projects proposed by his administration. (Odato, 1988)

Editorials in the county newspaper supported the incinerator at this time, but their tune would change in the heat of the ensuing struggle.

In response to an editorial endorsing the incinerator project, Binghamton's public works commissioner, Timothy M. Grippen—who was also running as the Democratic candidate for county executive at the time—published a "Guest Viewpoint" supporting members of Citizen Action of New York and RECYCLE who had called the existing county executive, Carl Young, "an opponent of recycling." Grippen reviewed what he referred to as "the record," showing that on March 4, 1988, Young said he strongly endorsed *voluntary citizen participation* in the recycling program but then changed his position after Grippen announced his candidacy for Broome County executive on April 15 "and stated very clearly that I planned to put a moratorium on incineration and immediately implement mandatory recycling." Grippen's article continued:

> On April 22, the *Press & Sun-Bulletin* reported: "Carl Young today is expected to call for a mandatory county-wide recycling program. . . ." Politicians watch public opinion polls very carefully and then try to be on the "right side" of an issue.
>
> Indeed, the *Press & Sun-Bulletin* poll on the subject showed that people in Broome County are overwhelmingly in favor of recycling. Not surprisingly, the county executive moved quickly to make pronouncements on the "right side" of the recycling issue in this election year. His actual commitment to recycling is, however, doubtful. In fact, his own recycling coordinator resigned in frustration over Young's foot-dragging on genuine recycling efforts.
>
> Broome County's solid waste program was put together with the help of consultants from Henningson, Dunham and Richardson (HDR). This company sold the county executive on incineration, claiming that the "new and improved" technology will burn unprocessed garbage at very high temperatures, which eliminates air pollution, and that the residual ash will not be toxic and can be safely disposed of in a landfill.
>
> HDR recommended only voluntary recycling to eliminate certain

> materials which do not burn. Like many non-technicians, Carl Young is too easily enamored of technology. Accepting HDR's arguments as fact, he bought their program wholesale. Now let's look at the facts. . . . (Grippen, 1988)

Grippen went on to document the existence of similar plants elsewhere—one of which also used HDR consultants—which were either shut down because of mechanical failures or unable to dispose of residual ash. He also mentioned the new 1988 air emission standards forced upon the federal Environmental Protection Agency and claimed there was no way the county or HDR could estimate the impact of those new standards. He continued:

> Given all the uncertainties, it makes sense to delay this project and re-evaluate the environmental and economical impacts when all the facts and costs are known.
>
> Citizens have a right to challenge public officials over public policy and they have a right to take their government to court if they don't agree. That's the beauty of this country's rule of law. We should salute the people of Citizen Action and RECYCLE for their actions on behalf of taxpayers.
>
> The county executive will claim any delay in the project will increase costs. He should know. He took a $15 million plan in 1981 and increased it to $40 million by 1986 and to $91 million in 1988.
>
> At this point, it makes good economic sense to at least consider downsizing the plant, and good environmental sense to delay incineration until after a full-fledged mandatory recycling program is in place.

Tim Grippen won the November election and over the next four years would work with the citizen groups to defeat this proposed incinerator.

Grippen's electoral victory galvanized the opposition. "Only in late 1988 did opponents of the Kirkwood site really become coordinated," according to Lee Cornell, materials recovery manager of Broome County in 1989. The former recycling coordinator mentioned above by Grippen, Cornell resigned in frustration from Young's administration after becoming convinced the county executive was not serious about recycling. "Kirkwood was the site selected because it had the least political opposition, . . . and not until Democrat Tim Grippen was elected county executive in 1988 did the opposition to that plant really start pulling together," Cornell told us. In contrast to the Philadelphia and Lackawanna County challenges, which started within a year of the first public mention of those proposed projects, this one in Broome County only really coalesced five years after the county's incinerator search had started, and more than two years after tentative identification of the specific Kirkwood site. Cornell continued:

> This incinerator project seems to have had strong political support in the county, until the election of 1988. It was run by a quasi-public body called the Resource Recovery Agency, supported by the political administration of Carl Young. . . . The Solid Waste Management Plan of 1988 identified a hierarchy for solid waste management involving waste reduction, waste reuse, recycling, then incineration, and then landfilling, and with this legislation they made fairly strong restrictions on the sizing and capacity of proposed incinerators in order to ensure that recycling had the best chance possible. . . . Now there is a question about whether the size of the proposed facility is appropriate for our Broome County waste stream.[6]

Grippen's campaign against Carl Young focused explicitly on halting this incinerator project, which Young strongly favored. Grippen insisted that serious recycling had yet to be tried. In the wake of the Grippen election victory in November 1988, Broome County legislators who favored plans to build an incinerator in Kirkwood adopted resolutions the following month intended to preserve the project by moving control for it from the newly elected executive, who opposed it, to the BCRRA, which was not under the new executive's control.

> Democratic lawmakers who support a proposed . . . trash incinerator are working to find a way to protect the project from their party colleague who is moving into the county executive's office in January. . . . County Executive Carl S. Young . . . is seeking legislators' support for a plan that would shift the Resource Recovery Project Manager . . . out of the executive's department and into the Resource Recovery Agency, a quasi-governmental body set up through state legislation to oversee the construction and operation of the incinerator. . . .
>
> Legislators said yesterday that their discussions with Grippen

6. Interview with Lee Cornell, Binghamton, New York, November 1, 1989.

> clearly showed there will be no compromise on the issue, and predicted an inevitable clash between the Democratic executive and the Democrat-controlled Legislature. . . . Legislator Albano E. Daniels, D-Endwell, the Legislature's representative on the Resource Recovery Agency, [said] "This is going to present a real problem. . . . Grippen's going to have a real fight on his hands. . . . I feel bad about it. I hate to start off with a confrontation, but for us [legislators supporting the incinerator] to do a 180-degree turn on this . . . I have to sleep at night." . . .
>
> Construction of the incinerator is scheduled to begin next summer, with work expected to take about 28 months. (Sbarra, 1988c)

"Grippen's election was a setback for incinerator proponents, but we still had a county legislature in favor of the project," Lori Dietz, BCRRA staff engineer, told us. "The national recycling debate, some NIMBYs, and local concern over health risks combined to promote this protest," she said. Dietz insisted that anyone interpreting Grippen's election victory as an anti-incinerator vote was simply mistaken:

> Grippen campaigned on two issues: opposing the incinerator and keeping taxes down. . . . I think he's totally misreading the results when he claims he really won on this incinerator issue. . . . The reason I say that is because at the same time this county executive election was going on, all nineteen legislators were also up for reelection. A number of them retired, so we have some freshmen legislators, but all that won did it supporting this project. . . . The main opponent against this project in the legislature lost her reelection bid.[7]

While sympathizing with opponents' hopes to see serious recycling reduce the Broome County trash stream, Dietz observed: "You can't build a waste-to-energy facility intended for the best possible recycling efforts." Her point was that if recycling results were under such estimates, the county would find itself with a new but undersized incinerator that was not capable of handling its actual volume of trash.

Whether to be interpreted as a mandate or not, Grippen's election on an anti-incineration ticket revealed that opposition to the trash plant was becoming something of a countywide phenomenon by late 1988, after Kirk-

7. Interview with Lori Dietz, Binghamton, New York, November 3, 1989.

wood officials had already signed the host community agreement with the vendor, Foster Wheeler, and submitted permit applications to the state DEC. The plus sign after the monetary compensation in Table 7 represents a number of additional incentives promised the host community.

> Back in September 1988, the county legislature approved an agreement with Kirkwood to compensate the town for hosting the county's incinerator by a vote of 16 to 3. Town Attorney Herb Kline then called it a "historic document." "No other community in the United States ever got as many different types of guarantees and incentives," he said.[8]

Grippen and other opponents argued, to the contrary, that the average Kirkwood resident had no input into town officials' decision to accept the project, and RECYCLE activist Audrey Glover echoed her neighbors' opinions in Kirkwood when she insisted in a personal interview that many viewed the agreement as having more negatives than positives for that community.

Although diverse in terms of background, attitude, and approach, the political opposition to the incinerator in Broome County was led by County Executive Tim Grippen. Organized citizens, such as the NYPIRG, Citizen Action, and RECYCLE, frequently worked with him in challenging the project. The three issues around which the opposition focused were the absence of prior serious and extensive recycling, reducing the size of any proposed waste-to-energy plant, and the question of where the ash from incineration would be disposed.

Just after Grippen took office, in January 1989, a RECYCLE lawsuit aimed at halting Broome County's incinerator project was thrown out of court because of technical errors on the part of RECYCLE's attorney and because the state Supreme Court judge insisted that he did not have jurisdiction in the case. While Grippen continued to oppose the facility, as chief county executive he was also obligated to honor the contract his predecessor had negotiated with Foster Wheeler. He publicly criticized the state Department of Environmental Conservation (DEC) for being committed blindly to incineration rather than to serious prior recycling and protecting the environment, but he admitted that he knew of no legal way out of the deal for Broome County. Grippen's alternative was to pressure the DEC from the beginning of his term to deny the incinerator's permit. His main argu-

8. Editorial, the *County Courier* (Broome County, N.Y.), February 6, 1991.

ments—elaborated in a long and detailed letter sent in March 1989 to the DEC, with a copy to New York Governor Mario Cuomo and other state officials, and also published in the newspaper—focused upon disposing of the incinerator's ash disposal; the large size of the incinerator, which he did not consider conducive to serious recycling; and unrealistically optimistic economic estimates by Foster Wheeler.

A permanent administrator of the Broome County Resource Recovery Agency, the agency responsible for the planned construction of the trash incinerator plant in Kirkwood, was named in May 1989. John Guinan, former Broome County deputy executive, was unanimously selected by employees of the BCRRA, despite criticism from those who insisted that the agency administrator should have a background in engineering. County Executive Grippen complained publicly that the BCRRA hired only former county employees who worked for former county executive Carl Young.

During the summer of 1989, Grippen caused a furor among some legislators when he said he would not enforce county laws and agreements requiring nonrecyclable garbage to be delivered to the planned incinerator. For example, a July 3 newspaper article said:

> In his most recent assault on the incinerator plan, Grippen said last week he has no intention of enforcing laws adopted last year to guarantee the delivery of garbage to the incinerator.
>
> A "flow control" law essentially gives county officials control over where all garbage in the county is dumped. That law, coupled with an intermunicipal agreement between the county and the agency overseeing construction of the burning plant, ensures that the planned incinerator would have fuel for its steam-powered electrical generators.
>
> The revenues of the burning plant will depend on money from electricity sales and on fees paid by haulers for dumping at the facility.
>
> By making it known he won't enforce the law, Grippen said he hopes to scare off investors who would buy bonds to finance construction, if not scare off Foster Wheeler. He said he would "rather go to jail" than see the county's garbage go to the incinerator. (Sbarra, 1989)

In late July 1989, DEC officials announced that the application for permits to build and operate the incinerator was complete and that the agency would move to the technical review of the project. At this time, some officials talked publicly of a spring 1990 date when construction might begin. Grippen continued to insist that serious mandatory recycling in the county would exceed the 40 percent level, leaving the county with less garbage than the minimum 130,000 tons a year promised in the Foster Wheeler contract. On the other side, BCRRA officials and some legislators insisted that no such level of recycling was possible.

DEC Law Judge Daniel E. Louis presided over the public hearings on the proposed incinerator in early November 1989. Residents of Broome County packed afternoon and evening sessions in a show of force geared to convince him that public opinion was against the project. A smaller number of project supporters argued strongly in its favor. Two observers in the audience told those of us sitting nearby that their daughter living near another plant in California, which Foster Wheeler claimed to be its safest plant, had mentioned in a telephone call the previous evening that the company only sent a taped message in response to nearby citizens' complaints.[9] After listening to arguments from both sides, Judge Louis was to consider them carefully and then relay his personal decisions and recommendations to the state's DEC commissioner, Thomas Jorling. Louis was under no specific time constraints, and Commissioner Jorling's decision would be final.

As is commonly the case with such capital intensive projects, the county newspaper officially supported the incinerator. One of its columnists, however, wrote numerous anti-incineration articles that opponents considered useful in their mobilization work. This columnist followed these public hearings and was critical of five legislators favoring the incinerator who had walked out of those hearings rather than listening to the various details that came out in the process.

> They do not want to risk having their minds cluttered with facts. They have been led down the garden path by Foster Wheeler and the Resource Recovery Agency. In the face of documented evidence to the contrary, they continue to believe that Foster Wheeler's plant in Commerce, California, is a towering success. So it can't get a permit to operate because it can't come close to meeting emission standards. So? Details, details.
>
> They are too polite or too gullible to demand that the Resource

9. From field notes taken at the Broome County hearings in Binghamton, New York, before Judge Louis, November 2, 1989.

> Recovery Agency produce a bid evaluation for the incinerator job. . . .
>
> In the waning days of 1988, the legislators snatched the Resource Recovery Agency out from under the jurisdiction of the county executive—where they were quite content to see it lie when Carl Young was boss—and established it as an autonomous, quasi-governmental body.
>
> Ask and ye shall receive, the legislators told the RRA people. They did and they did. For openers, the Legislature handed over $685,000 in cash. Then it gave the agency all the county-owned desks, chairs, and other furniture it had used when it was part of county government, along with telephones and a computer system. . . . What has the agency done with that $685,000, you ask? Well, since the agency has become a sinecure for casualties of Grippen's 1988 election, including the erstwhile county attorney and the former assistant county executive, a good part of it has gone to pay their salaries.
>
> Now, with its till empty and having accomplished nothing tangible, the agency has come back to its legislative benefactors rattling the cup and asking for another $277,000 to tide it over. . . . Rest assured the fiscal watchdogs in the Legislature will see that the agency gets its money. . . . (Rossie, 1989)

As predicted, the BCRRA received the money, but support for the project weakened as the county legislature barely raised the twelve votes needed to override Grippen's veto of the $200,000 set aside in the 1990 budget for the agency.

The controversy intensified in 1990 as opponents pressured the DEC to withhold the permit, and the BCRRA pushed for a positive response from the same state agency. As researchers, we found ourselves unintentionally involved to a minor degree. Broome County activists requested the results of our 1990 backyard telephone survey, which differed from a broader survey of county residents' attitudes toward the disposal of solid waste commissioned by Foster Wheeler revealing 63 percent of the people favoring construction of the proposed facility.[10] Only 10 percent of those in our survey said they favored the siting. Not wanting to change roles from "observers" to "players," and yet recognizing the legitimacy of the activists' request,

10. The survey for Foster Wheeler had 401 respondents and was conducted by Yankelovich, Skelly and White/Clancy, and Shulman.

we agreed to wait until we had received the mail questionnaire responses from the activists and CAC members, before sending the Broome County telephone survey results to both Audrey Glover of RECYCLE and Lori Dietz of the BCRRA months later, while the controversy was still in process. Although our telephone survey was referred to in subsequent hearings, it did not figure prominently in the conflict.

DEC Commissioner Jorling came to Binghamton to preside over rare supplemental hearings in mid-August 1990. These were precipitated by the intense and sometimes bitter struggle among county officials over the appropriateness of the proposed plant.

> Jorling said his presence at such a hearing is unusual, but that the controversy in Broome County calls for "extraordinary steps." In an interview, he said the situation is extraordinary because the matter has so divided county leaders. . . . Jorling said: "You've got a legal situation in Broome County where you've got a county Legislature with a viewpoint, a county executive with a viewpoint, and an agency with a viewpoint."
>
> The differences were apparent Wednesday, when, after Grippen gave a 20-minute argument against the project, Legislator Albano E. Daniels, D-Endwell, urged the commissioner to "ignore the disgruntled wailings of the county executive and issue the permit." Daniels represents the Legislature on the Resource Recovery Agency. . . .
>
> The key disagreement is over how efficiently Broome County can recycle and reduce its waste. Foster Wheeler and Resource Recovery Agency officials say 165,000 tons a year will be available for burning; project opponents say less than 130,000 tons will be left after sorting out recyclable and non-burnable refuse. By its contract with Foster Wheeler, the county must provide at least 130,000 tons a year or pay a penalty. Last year, about 217,000 tons of garbage were buried at the county landfill.
>
> How much garbage is available to burn determines how much county users will pay per ton of trash. Economic analyses conducted for the Resource Recovery Agency show, for example, that if 165,000 tons a year were available, the per-ton price in the first year would be $36, compared with $66 a ton if there were 117,600 tons to burn. . . .
>
> Many angry Broome residents told Jorling Wednesday night that

> a recycling program should be given a chance to work before the incinerator is chosen as a waste disposal solution. (Lau, 1990)

The following month, Jorling ruled that the incinerator permit would not be issued until four major issues were resolved, after being reviewed before an administrative law judge in an "adjudicatory hearing." These issues involved size and economics, the county recycling plan, ash disposal, and air quality. Then in October 1990, the Broome legislature voted 11–8 to cancel the county's contract with Foster Wheeler, although that was actually never done, as explained below. In November, Judge Louis ruled that no hearings would be held until the matter of where to bury the ash residue from the incinerator was settled and the BCRRA could show that it had at least five years' worth of state-permitted landfill space for the ash (Sbarra, 1990).

Because of the increasing opposition, the Broome County Legislature voted 10–9, on December 4, 1990, to recommend to the BCRRA that it terminate its contract with the incinerator vendor, and at the same meeting the Legislature also voted 13–6 against a resolution that would have allowed bonds to be issued to support the project. Less than a week later, however, six members of the BCRRA signed a letter sent to County Executive Grippen informing him they would not terminate the contract, despite the Legislature's vote, and that they would "defer to the 1991–92 Legislature for direction."[11] Despite the BCRRA's confidence that the newly elected legislators would be more favorable, Grippen insisted, in the same letter, that theirs would not be a final verdict:

> Even if the new Legislature does change its policy relative to the incinerator, the project cannot move forward without the Commissioner of DEC issuing a permit, which he has ruled he will not do, until such time as Foster Wheeler and the Agency provide for ash; determine the best available control technology and the proper size of the facility; and finally, allow the "community" an opportunity to decide if the cost of the project is reasonable given alternative methods of disposal.

11. Grippen explained this in a long letter, January 15, 1991, to Audrey Glover, a Broome County activist helping lead the opposition to the Kirkwood siting. Her generous assistance in reconstructing the story of the Broome County protest—including a sharing of this letter from County Executive Grippen—has been critical in this project.

The new County Legislators voted 10–9, in early 1991, to fund the BCRRA, but Grippen vetoed their decision and explained that he was searching for a way to force the Broome County Legislature to another vote on the incinerator, because he was convinced it would be a negative vote this time. In a show of strength by the new Republican majority, the Legislature voted 12–7 in February 1991 to override Grippen's veto of a resolution to fund the BCRRA.

> Legislature Majority Leader Emil J. Bielecki, R-Vestal, took the lead in marshaling support for the override. . . . He lashed out at Grippen for his lack of leadership. . . . Legislature Chairman Arthur J. Shafer, R-Kirkwood, a longtime opponent of the project, also supported the override. Citing a letter signed by a majority of Kirkwood board members urging him to protect the town's agreements with the county, Shafer said he was "going to set aside my personal views and act on behalf of my constituents."
>
> Audrey Glover . . . said Shafer is not representing the people who live in the three towns in his district, but the Kirkwood Town Board. "This is irresponsible government," Glover said. . . .
>
> "Unfortunately, it comes down to the Legislature just giving up any possible responsibility they have for the project. The agency will now make decisions regarding this project," Grippen said.
>
> Opponents fear the DEC hearing process will produce the permits needed to build the incinerator. . . . The final DEC hearings on the project are expected to start about April or May. A final decision is not expected until about August. Construction is planned for as early as January to be completed in early 1993. (Sbarra, 1991a)

After suggesting that the Legislature's vote "probably established the political ground rules for the next two years" in overriding Grippen, the county newspaper's editorial board made it clear that the *Press & Sun-Bulletin*'s position on the incinerator had changed from supporting it to opposing it, as mirrored in an editorial of February 25, 1991:

> Obviously, the Republican leadership decided it was necessary to teach Grippen a lesson, even if that meant doing a disservice to the county by keeping alive a flawed project and further delaying the adoption of a comprehensive solid waste plan. . . .
>
> The Resource Recovery Agency is a one-track operation, with an

> $80 million, 570-ton-a-day incinerator at the end of the line. . . . Incineration has a definite place in a comprehensive solid waste management plan, but this particular incinerator doesn't meet Broome County's specific needs. Nor does it constitute a responsible plan.

Some Broome County opponents traveled to the state capital at Albany in early March 1991 to pressure Governor Cuomo to influence DEC head Jorling to rule against the Broome incinerator project. They were among the approximately 400 people who marched from the Capitol to the Executive Mansion to urge the administration to spend more on recycling programs and to denounce garbage incinerators. The rally was organized by the NYPIRG and Work on Waste (WOW), a group discussed later in this chapter. The rally and march were the first phase of a two-pronged campaign that also involved lobbying the state legislature on incineration issues. Protesting in front of the Executive Mansion, activists challenged the governor to imitate his New Jersey counterpart, Governor James Florio, who had recently turned his state from an emphasis on incineration to an emphasis on recycling, chanting "Hey, ho, Mario . . . Be more like Florio!" NYPIRG attorney Larry Shapiro told the crowd: "New York burns more garbage than any other state. . . . We need to tell the New York State leaders to end their love affair with incineration":

> Commenting on a state law that requires municipalities to develop recycling plans by 1992, Shapiro said the state must help foot the bill. "By not coming up with recycling funds," he said, "the state is not holding up its end of the bargain."
>
> [Queens College Professor Barry] Commoner said the state needs only to look a few miles south for inspiration. "Right now, New Jersey is recycling at a 40 percent rate," he said. "That state had a plan that began with a moratorium on incinerators. Then, they aimed for a 60 percent recycling rate. . . . New York doesn't think you can recycle."
>
> Commoner also criticized the government for ignoring a study he conducted on Long Island that he said showed that "84 percent of household garbage can be recycled." It was a study for the state—"They've got the report," he said. (Roy, 1991)

After the BCRRA had located a Virginia firm to dispose of the ash and unburnable waste from the proposed incinerator, as stipulated by the ruling of Judge Louis in late 1990, the final DEC hearing on the project started in June 1991. Agency Director John Guinan publicly expressed his confidence that actual construction might begin in October of the same year. Although original plans for the incinerator included provisions for the ash residue to be buried at Broome County's Nanticoke landfill, difficulties in obtaining the DEC permits to keep that landfill open complicated the issue. This DEC hearing would include discussions concerning some of the most salient topics in the incineration debate, focusing especially on four key areas:

The size and cost of the facility
The equipment needed to adequately control toxic air emissions
Provisions for disposing of the ash residue from the plant
The conditions necessary to ensure that burning won't inhibit county efforts for meeting the state's 40 percent recycling goal

In his testimony, Grippen claimed that "the most critical issue of all is the sizing question—that will drive what happens. The facility is twice as large as we need and it will cost about twice as much as we need to pay," suggesting that incinerator operators be required to take the politically unpopular step of looking for outside garbage to import "to keep the fires burning efficiently" (Sbarra, 1991b).

Incinerator proponents viewed things differently, and the formal hearings did little to change opinions that had crystallized over almost a decade of conflict. At their conclusion, Foster Wheeler's attorney said, "It's always hard to say what the commissioner will do, but I would be surprised if he didn't issue a permit. . . . I think we've done everything anyone could ask of us." The attorney representing several citizen groups fighting the incinerator, on the other hand, observed: "If [DEC Commissioner] Tom Jorling issues a permit for this incinerator on this record, then Tom Jorling is environmental enemy No. 1 in New York State. . . . On this record, Tom Jorling has no rational basis to issue a permit." The majority of observers had more qualified views:

> Most officials . . . said Jorling will likely at least limit the amount of garbage that could be burned. Such an action would be in keeping with DEC staff testimony recommending the facility be limited to burning no more than 125,000 tons of garbage per year, instead of the 165,000 tons used in planning the project. That would mean

> higher fees to residents using the facility, with an increase to about $70 per ton from about $50 per ton.
>
> DEC Regional Attorney Richard J. Brickwedde said the tonnage limit was recommended by DEC staff to maximize county recycling efforts, as Jorling wishes. He said the staff is not recommending a smaller plant because it's possible the county won't be able to recycle the amount the state hopes and may have to burn more. (Sbarra, 1991c)

While waiting for Jorling's decision—initially due in October 1991—incinerator opponents worked to popularize a perception of recycling as the alternative to incineration through public meetings and letters-to-the-editor, mobilizing the grassroots, and criticizing specific BCRRA proposals. Some county legislators who had earlier supported the incinerator talked publicly about having second thoughts after the DEC hearings, where they realized that tipping fees were likely to jump from a projected $51 per ton to $70 per ton or more because of Jorling's insistence on serious county recycling before incineration. Other legislators suggested it might be time to consider Broome's a regional plant, which would import garbage from neighboring counties to keep these fees down. That was not a popular idea with many county residents.

Opponents worked at both the legal level and the political level. Attorneys for the NYPIRG and RECYCLE kept the pressure on by filing suit in July 1991 against the BCRRA for not conducting competitive bidding before signing a contract to dispose of the proposed incinerator's ash in Virginia. RECYCLE's Audrey Glover, reflecting on her expectations regarding the pending decision by Jorling in a September telephone interview, observed, in part:

> The best we can hope for from the DEC commissioner is probably a size reduction. . . . The county is really dragging its feet on recycling. . . . Citizen groups prefer serious trash reduction, recycling, and safe landfilling, but the incinerator proponents really sized this plant for the region rather than our county from the start.[12]

NYPIRG and RECYCLE activists also staged a rally in late October involving approximately 120 people, outside Governor Cuomo's Binghamton of-

12. Telephone interview with Audrey Glover, September 28, 1991.

fice, against the proposed Broome incinerator. RECYCLE Chair John Smigelski told listeners at the rally that "Foster Wheeler testified under oath that for two years they knew they couldn't meet the permit conditions for mercury and they had lied in their permit application. The DEC staff said nothing." An account of the rally continued:

> "The proposed incinerator in Broome County would pollute the air with lead, mercury, dioxins, and any number of other poisons," said Larry Shapiro, [NYPIRG] staff attorney. "It's enormously expensive. Broome County is the place where the incinerator industry's efforts to turn the lungs of New Yorkers into toxic waste dumps must be defeated. The Cuomo administration has never denied an incineration permit," Shapiro continued. "The people of Broome County have offered Governor Cuomo a rare opportunity to become an environmental hero. He has a chance to stop incineration forever in New York State." (Musselwhite, 1991)

Arguing that the Republican majority in the Broome County Legislature viewed recycling as an enemy of incineration, the associate editor of the county newspaper wrote:

> How, you may ask, could any responsible citizen, in this environmentally aware age, not be enthusiastic about recycling, which has, in some circles, taken on near religious significance? The answer is simple: When recycling gets in the way of a higher commitment. And the higher commitment in this case is the oversized, overpriced incinerator. . . .
>
> Successful recycling means less waste to burn, and that spells trouble for Foster Wheeler, the RRA and its tame legislators, and their dreams of a regional trash burner.
>
> Data developed by the Broome Environmental Management Council and the League of Women Voters show that the tonnage of waste received at the Nanticoke Landfill has been cut in half, from a high of 228,000 tons in 1986, to approximately 125,000 tons so far in 1991. Recycling is principally responsible for that, and keep in mind that recycling in this county is far from full-scale. . . .
>
> Recycling, the hope of the environment, is the enemy of the RRA's incineration plans as they are now construed and all those who are pushing those plans. (Rossie, 1991)

The long-awaited ruling came from Jorling's office on December 18, 1991, surprising both sides. The DEC commissioner gave applicants thirty days to either propose a smaller plant or provide binding agreements to take between 35,000 and 40,000 tons per year of waste from outside Broome County.

> Jorling, in an 18-page decision, said he believes Broome will have only 114,000 tons of burnable waste a year by 1997, not the 165,000 estimated by the applicants.
>
> Foster Wheeler Corp., the company contracted to build and run the $80 million incinerator, remains committed to the job, said John Guinan, administrator of the county Resource Recovery Agency, the agency overseeing the project. . . .
>
> Jorling said the proposed incinerator is oversized by about 45 percent. "The decision issued today establishes how such facilities must be sized to complement, not conflict with, waste reduction and recycling efforts, which must continue to be the cornerstone of our solid waste management strategy," Jorling said in a prepared statement issued with his decision.
>
> County Executive Timothy M. Grippen, who has opposed the incinerator proposal since before he took office three years ago, said the conditions Jorling imposes are what his administration has long sought. . . . Grippen said that while he accepts the concept of a smaller plant, he would have to see whether one would be financially practical. . . .
>
> Since Jorling left open the incinerator question, project opponents said they were heartened by the decision, but not declaring victory. (Lau, 1991)

This ruling, in essence, said that the county did not produce enough waste to fuel the proposed 570 ton-per-day burner while still meeting New York's recycling and waste-reduction goals. In a December 20, 1991, article, the county's *Press & Sun-Bulletin* reported that nine Broome County legislators said they would support importing trash, eight said they would oppose, and two were undecided. Grippen said he would veto any effort to import trash but would not oppose a plan to reduce the size of the facility. A series of local television editorials running from January 10 to January 15 on WBNG (Binghamton) summarized the struggle and the "downsize or import" options and concluded by supporting the former: "Build a smaller, downsized plant, move ahead with aggressive recycling, and get on with the rest of the business of government."[13]

The Foster Wheeler Corporation expressed no interest in building a smaller incinerator; the vendor and its supporters favored the importation of trash from outside Broome County. Opponents argued that this would make the county a regional dump, distributing flyers with legislators' telephone numbers and urging people to pressure them to "stop importation and the incinerator." A special session of the Broome County Legislature met in early February to decide whether to import garbage and continue funding the BCRRA. In what some called a "veto-proof" vote, they opted to import trash to save the incinerator plans.

> Broome County's plans for an $80 million trash incinerator were pulled from the ashes Thursday by a legislative vote to import garbage. A measure that would allow the county to seek up to 40,000 tons of trash per year from neighboring counties was approved by the county Legislature, 12–7, the minimum needed to survive a promised veto by . . . Grippen. (Sbarra, 1992a)

As of early February 1992, even the organized opponents were thus resigned to the likelihood that an incinerator would be built after a positive vote by the Legislature to import outside trash. As promised, Grippen vetoed the plan, but in late February, contrary to all expectations, one legislator changed his override vote at the last minute, alleging that his request for more time to study the possibility of a public referendum was ignored by those forcing an immediate decision.

> Garbage will not be imported into Broome County for a proposed trash incinerator, county legislators decided unexpectedly in a vote last night.
>
> The legislature voted 11 to 8 in favor of trash importation, one vote short of the 12 needed to override . . . Grippen's veto last Friday. . . . Many legislators and other public leaders expected the decision to pass, because 12 legislators had voted in the original vote to import trash.

13. John S. Mucha, vice-president and general manager of WBNG-TV, Binghamton, New York, "Broome County's Waste-to-Energy Facility, Part IV" (television editorial), January 15, 1992.

> Daniel A. Schofield (R-Endicott), who voted for importation in the original vote, provided the critical "no" vote last night. He said he voted as he did after the assembly denied his request that a vote not be taken until the assembly had considered all options. He said he felt the assembly was rushing into a decision to import garbage. . . . "I pleaded with the assembly not to make the decision on this veto matter last night," he said. . . .
>
> Last night's decision means that in order for incineration to be considered again by the county assembly, an entirely new resolution would need to be brought before them. "As far as I'm concerned, incineration is done in Broome County," said Arthur Shafer, chairman of the legislature, at the conclusion of last night's meeting. (Apter, 1992)

As things turned out, this vote was the death of the Broome County incinerator. Foster Wheeler warned the county that it had an obligation "to pursue plans for building a smaller version of a proposed 571-ton-per-day incinerator," even though the same company had earlier insisted that "a smaller plant would require an unacceptably high user fee." Legislature Chairman Shafer criticized Foster Wheeler for this action, noting that "the reality is that they were the ones who argued with me that a down-sized facility just wasn't feasible" (Sbarra, 1992b).

The BCRRA closed its doors in late February 1992, and county officials eventually focused on composting and the continuing use of the landfill for handling the county's trash. Foster Wheeler filed a lawsuit against Broome County, Grippen, and the BCRRA in February 1993, prompting one legislator, Majority Leader Louis P. Augostini, who supported the incinerator plan, to comment that the suit would likely accomplish "what the incinerator proposal never did—bring county leaders together. . . . We're all on the same side now." A newspaper report added: "Whether you were for or against the incinerator, that's got to be put aside. It's time to defend the county" (Sbarra, 1993a).

The Broome County Solid Waste Division's year-end report for 1992, released in May 1993, showed that the county "recycled, composted, reduced or otherwise avoided burying about 49 percent of the 282,925 tons of garbage county residents had been expected to generate" (Sbarra, 1993b).

Before turning to the last siting conflict, farther north in the same state, it is instructive to see what critical junctures opposition activists and CAC members identified in this multiyear process. Thirty of the thirty-four questionnaires mailed out to Broome County activists were completed and returned. Each named RECYCLE, Citizen Action, and/or the NYPIRG as his or her primary protest group, which eventually coordinated activities within the loosely knit "Don't Burn Broome" coalition. These activists generally agreed on four major turning points in the conflict and believed that they had played important roles in each: the election of Grippen, the formation of their countywide coalition, focusing opposition primarily upon economics, and mobilizing pressures on elected representatives. Some illustrative responses to the question about main "turning points" in the development of organized protest: "winning the county executive election," "developing a Don't Burn Broome coalition among RECYCLE/Citizen Action/NYPIRG," "focusing on economics instead of environmental issues," and "our public pestering of Governor Cuomo on his visit to Broome so that he sent DEC Commissioner Jorling for an unprecedented extra public hearing on the incinerator." Of the responses to the related item inquiring about "decisions by the group or its leaders which turned out to be very important," the following are typical of those most frequently mentioned: "decision to apply for party status at permit hearings," "focusing on economics and imported garbage," "pushing for a vote to cancel the project before 1990 elections put serious pressure on a number of legislators who changed their positions," "standing our ground at the public meeting in the park until Governor Cuomo came over and heard us out," "focusing activity on a target group of county legislators—pressuring and punishing them until they voted our way."

The leader of the local protest group closest to the proposed site, John Smigelski of RECYCLE, appended the following insightful comments to the end of his questionnaire, returned in November 1990 before a final decision on the incinerator had been made:

> My weakness was in organizing skills and not being able to keep a group growing. My strengths were energy and technical skills—engineering, MBA, and knowledge of laws and licensing procedures. . . . It took us a while to learn how to use the media, but we eventually did. RECYCLE did not grow into a countywide group—we remained primarily local as far as active members were concerned. I had a personal friend at Citizen Action who brought us skill training and a forum to go countywide. After a while we formed the Don't Burn Broome Coalition.
>
> I wound up providing a lot of the technical analysis and training

> for the other coalition leaders as we developed strategy. The decision most groups face is whether to focus on legal or political actions, and we went back and forth on this issue. The legal approach can provide delays that may kill the siting process, but only politics can really kill a project. We tried to do everything, but the legal was a mixed blessing for us: the timing of our lawsuit helped Grippen get elected the new county executive, but our failure to win it gave us a "loser" image. That's why the coalition took over the public effort.
>
> We focused on town boards and organizations, emphasizing size and economics. A big turning point was the large public turnout when Mario Cuomo visited. Many believe that was key in helping us scuttle the permitting process.

Responses from six of the nine CAC members who received questionnaires supported our earlier suggestion that this group played a minor role in the Broome County struggle, and even that was not always on the proponents' side. For example, in a questionnaire item asking about very important CAC decisions, one wrote: "[Some of us] concluded that the plant was much too large and would require the importation of garbage to maintain its operation." Another said the CAC had encouraged "several minor changes and additions to the health risk assessment" but had really done "nothing very important." A third explained why he and others on the CAC concentrated more on monitoring than on eliminating the project: "Because legal and financial consultants advised us that a host community had never won a suit preventing a higher government body from siting in their community, we worked to achieve the highest level of control over plant emissions."

Two longer comments written at the end of these CAC members' questionnaires provide additional insights into the Broome process. The first noted:

> Unfortunately, it was decided early on that recycling could not work and that the best solution was to simply burn the majority of the waste. In the county's defense, this seemed to be the prevailing attitude across the nation. This led to the very frustrating experience of going through the motions of public input (sanctioned by the county) without any corresponding influence. I also feel that once the county administration made the decision to build such an expensive project (this was to be the largest public works project ever undertaken by

> the county), monied interests pushed the project, rallying both the corruptible and the gullible. No rational decision was to change the overall strategy.

The second comment emphasizes the importance of changing perspectives:

> When the incinerator was proposed in 1980, people were not very interested in recycling. As the 80s progressed, the cost—social and monetary—of landfilling increased and willingness to take personal responsibility for decreasing the waste stream did also. Until then, incineration, as bad as it is, made more sense than landfilling and therefore I couldn't really oppose it. Now, however, the times are different and the incinerator seems like a poor solution. As political science tells us, long-term rational planning (like the 10 year process for the incinerator) seldom works well in the policy process because policy isn't made this way and the problems change. Incineration is like nuclear power in this regard.

The St. Lawrence County Siting Attempt

Another siting effort in New York—this one at the northern tip of the state in St. Lawrence County—had also been initiated in the early 1980s by a county legislature.[14] As in Broome, a special organization to address solid waste problems was created and worked for approximately a decade to construct an incinerator. Associated with this project was also a nominal citizen advisory committee, which was similarly defeated by a surprising last-minute vote in its own county legislature. Perhaps the most dramatic differences between these two siting attempts derive from characteristics of their challengers. A citizen group outside the target town of Ogdensburg played a major role in mobilizing anti-incineration forces against the project. This social movement organization, approximately seventeen miles away, called itself WOW—for "Work on Waste"—and provided the base from which

14. The following individuals—both proponents and opponents of this particular project—offered useful insights about the St. Lawrence County siting attempt: Mary Verlaque, Ellen Connett, Ruth Beebe, Paul Connett, Klaus Proemm, Richard Grover, Keith Zimmerman, Tom Plastino, and William Sutkus.

Paul Connett, a faculty member in the chemistry department of St. Lawrence University in Canton, emerged to become a national leader in the movement against incineration.

In the early 1980s, the St. Lawrence Solid Waste Disposal Authority (SWDA) was formed by the St. Lawrence County legislature to address the problems generated by landfill scarcity. Few people opposed incineration at this time, so Richard Grover happened to be a lone crusader, and he would eventually contribute to Paul Connett's transformation into a pro-recycling and anti-incineration activist. Grover was the director of planning in St. Lawrence County from 1970 to 1979. He had received his degree in planning and landscape architecture from the University of Pennsylvania in the 1960s and had worked previously in Ohio and Cape May, New Jersey. In the early 1970s he had helped direct a study for St. Lawrence County that concluded: "If you can't recycle it, you shouldn't make it."[15] In 1978, while still director of planning, he worked to develop a comprehensive land use program that was adopted by the planning board and accepted by the county legislature and even by educational institutions. "It got rave reviews," Grover noted, "but then my demise as director of planning came when I was busted for marijuana possession in the spring of 1979 and, although the charges were dropped, I had to step down."

Before the above incident, Grover had hired Mary Verlaque for the position of, as he put it, "Number 2 Planner in the County." Verlaque took over Grover's position—becoming the first acting director of the SWDA after that organization was created in 1983—at a time when incinerator companies were advertising themselves as the answer to the landfill shortage. She surprised her former supervisor, Grover, by gradually turning wholeheartedly to burning trash, becoming, in his words, "the prime sponsor of the incinerator, who won the planning board over to her side." A former county official well acquainted with the history of the incinerator struggle, Tom Plastino, explained that the legal structure of industrial revenue bonds made incinerators a significantly more favorable investment in the early 1980s than after 1987.[16] Mary Verlaque, still the director of county planning in

15. Taped interview with Richard Grover, Canton, New York, December 12, 1990. This informative two-hour discussion is the source of direct quotations from Grover throughout the following pages unless otherwise indicated.

16. Tom Plastino, formerly of the St. Lawrence County EMC and later director of the SWDA staff, read over an earlier draft of this chapter and offered numerous helpful comments in a letter to the authors dated November 18, 1993, and in a June 3, 1994, telephone conversation with Ed Walsh.

September 1990 after the legislature vote had gone against bonding for the incinerator, was nevertheless shocked at the way things turned out: "Once the opponents started instilling doubts, the public kept slipping away," she observed. "None of us expected the legislators to take the protesters so seriously."[17]

Initial anti-incineration activists had a credibility problem in the mid-1980s, when both the New York State DEC and county officials were enthusiastically endorsing that technology. The chastened Grover, for example, was a voice in the wilderness, and although a few others eventually joined him, none of these early activists were from the political mainstream—as was the case, for example, with the newly elected Broome County executive. Grover noted the problem:

> The state of New York was really setting the legislators' agenda, because it was a government without a policy, which had money to spend on solid waste management. . . . Most of their money was going for incinerators, partly because incinerators were the only thing out there at the time. . . . The New York State DEC was and is heavily laden with engineers who think "burn" rather than "recycle" because it's not even in their frame of reference. Engineering schools teach people how to build things, and recycling isn't building things. . . .
>
> My standing in the community was pretty thin. . . . I was just beginning the road to political recovery. It's been a long haul. . . .
>
> This is probably a good point to tell how Paul Connett fits into this whole thing. He's now a heavyweight in fighting incinerators around the globe, but I introduced Paul to this issue while I was teaching a course at St. Lawrence University part time, picking up a few bucks here and there after I lost my job. I was teaching land-use planning, environmentalism, and related things to support myself. . . . That's where I met Paul in '82 or '83. . . . He didn't know the first thing about garbage—a neophyte as far as this whole field was concerned. . . . What's relevant here is that Paul is British, and he became the leading spokesperson for WOW and the opposition.
>
> Who did you have here articulating a no-incinerator platform? You had the deposed director of county planning and this college

17. Telephone interview with Mary Verlaque, September 27, 1990. Any observations or quotations of Verlaque's not otherwise identified in the following paragraphs derive from this interview.

> professor with a British accent! The mainstreamers asked with scorn, "Who are these guys to be telling us what to do with our garbage and how to run our county?"
>
> So personalities definitely enter into this—we were clearly outsiders. Paul has a Monty Python style, a real showman, and he gets up at the county legislature and really blasts them. And they weren't about to be moved one iota by these renegades.

Paul Connett became actively involved in this local siting process during 1985 when he contacted Barry Commoner and joined him in challenging Floyd Hasselriis's data on burning dioxins (see Chapter 1). Connett was disturbed when members of his own county's planning board seemed unimpressed by what he regarded as solid scientific challenges to Hasselriis and contented themselves with explaining away inconsistencies in his data with such comments as "Well, I've met recyclers who lie!"

In the autumn of 1985, Connett received support from his university to attend an international symposium on dioxin in what was then West Germany, and with Barry Commoner he became increasingly involved in challenging incinerator projects throughout New England. By the summer of 1986 he was at the center of a loose-knit network of anti-incineration activists across the United States, but that was not necessarily to Connett's advantage as he argued against the nearby Ogdensburg incinerator. Observers noted the county's "characteristic nativism," which very much resented "outsiders trying to tell St. Lawrence County people what to do."[18]

Observers on both sides agree that 1986 and 1987 were pivotal years in the struggle. While Connett's anti-incineration group—Work on Waste, or WOW—played a central role in leading the local grassroots challenge during the latter 1980s, it probably could not have stopped this project without assistance from other sources. For example, not only did the St. Lawrence SWDA make a critical change of incinerator vendors in December 1986, but the state of New York also issued new regulations emphasizing recycling in 1987. In addition, a local trash hauling firm with which Richard Grover was associated by this time, Waste Stream Management (WSM), was also emphasizing recycling and making significant demands on the SWDA that worked to WOW's advantage. Grover explained some of these developments:

18. While these quotations are from the previously mentioned comments of Tom Plastino (note 16), they are typical of what was frequently said about St. Lawrence County residents' attitudes toward outsiders.

What happened in December '86 is an important part of this story because SWDA had been going through a process to select a vendor that started in the early 1980s. . . . It's a long and complicated process. You issue an RFP, get proposals from vendors, select a vendor, and you enter into a contract, and then the vendor carries out all the environmental studies necessary for securing a permit. . . . What happened in this particular case was that the St. Lawrence County SWDA could not come to terms with the vendor, Sigoure-Freres, because this French firm did not have the financial stability the SWDA was calling for. . . . In December '86 SWDA terminated its agreement with Sigoure-Freres, and sought another vendor. That turned out to be a fatal development because they had to start the whole environmental review process over. . . .

The county's new draft RFP [Request for Proposal] came out in February '87, but so did the New York State solid waste plan that set recycling and reuse as top priorities. . . . It was a new ballgame. . . . SWDA did not anticipate the role of recycling in this new scheme of things, and so did not change the design of their project when they put out the new RFP. In fact, they actually made the incinerator bigger! Went from 225 tpd to 250 tpd—flying in the face of the real political world which was emphasizing burning less rather than more! . . .

Then, in June '87, Waste Stream Management, a big hauler/recycler in the county, proposed a massive recycling and composting effort that the SWDA turned down, claiming it had to control the entire solid waste management output of the county in order to guarantee the financial stability of the incinerator project. . . .

Because they knew they were dead meat in the eyes of the state regulators if they admitted this, they couldn't come right out and say "We are going to burn recyclables in the incinerator." So they played an extremely dishonest numbers game, using their consultants out of Falls Church [Virginia] by the name of Gershman, Brickner, and Bratton. They falsified the blueprints, saying the county's going to grow after the incinerator comes on line in 1992, but the county was not actually growing at all, and is really more likely to continue losing population than to grow in the future. . . .

So my point is that the waste generation data were manipulated by the consultants to attempt to show that it was financially feasible to recycle and build the size of incinerator they wanted. . . . A major

> point was that recycling was going to get phased in over something like a ten-year period to support only the state-mandated minimums for recycling. . . . So they manipulated the population and waste generation numbers to show there was still enough garbage available to make this project work without importing garbage, because they didn't want to fight the political battle that importation was certain to cause.

Waste Stream Management's lawsuit against the St. Lawrence Solid Waste Disposal Authority during this period was a major factor in delaying the project. Emphasizing the independent importance of this litigation for the eventual defeat of the incinerator project, former county official Tom Plastino observed:

> It is essential to understand that no number of WOW bake sales would have been sufficient to postpone the [incinerator] juggernaut until after the 1989 elections had not WSM been prepared to spend big bucks to fight the SWDA both in the courts and in the public venue.
>
> Few people seem to realize how unusual this was. Here's a company in the middle of nowhere saying—and maybe even believing for a few years—that its economic future lies in recycling, not just in hauling trash to burn or bury. . . . They spent, I would guess, several hundred thousand dollars, counting the principals' time as well as cash, to fight the project to the bitter end. Had WSM not had the money and the will to do so, the incinerator would have been under construction long before November 1989 and there was nothing that WOW or the economic "realists" would have been able to do about it.

The citizen protest group WOW started holding weekly meetings in 1987, a practice that would continue for the next three years. Based in Canton, this grassroots organization worked to replace pro-incineration incumbents with anti-incineration candidates in 1989. Ruth Beebe, a WOW activist, noted that "until 1988 every vote by legislators was unanimously pro-incinerator, but last November [1989], many of those legislators were ousted with considerable help from WOW."[19] Another WOW activist, Klaus Pro-

19. Taped interview with Ruth Beebe, Canton, New York, December 11, 1990.

emm, who worked for the State Employment Office, explained why the proposed host city, Ogdensburg, which would receive some financial incentives, had more reason to favor the project than Lisbon or other municipalities, such as Canton or Potsdam, where opposition was strongest:

> The state of New York encourages regionalism whereby one county imports another's waste, but none of us want to be the receivers of outside wastes. St. Lawrence County generates between 150 and 160 tons per day, with an absolute maximum of 200 tpd, and yet the proposed incinerator was to burn 250 tpd, so it was to handle all this county's garbage plus more. . . . Ogdensburg is almost completely under the influence of the "powers that be," but you can understand those residents' reasoning, because the county would be paying them. Lisbon, however, was downwind from the proposed incinerator and also scheduled to receive the ash, so it would be absorbing the most impact.[20]

It is not surprising, in the light of this analysis, that WOW was more successful in its anti-incineration efforts in Lisbon than in Ogdensburg. Proemm explained that a number of businesspeople in the Lisbon area would eventually join in opposing the proposed plant "near the end of our struggle because they saw the incinerator as a bad business deal." But this is getting ahead of the story.

After terminating its contract with Sigoure-Freres in December 1986, the SWDA issued another Request for Proposal, which eventually led to a new agreement with Harbert/Triga Resource Recovery Inc. of Birmingham, Alabama, on July 19, 1988. This was to be a 250 ton-per-day incinerator costing approximately $25 million, but the SWDA would suffer major credibility problems when, a few months later, a confidential draft of an embarrassing risk assessment document was leaked to the press.

Storm clouds continued to form over the proposed incinerator in the early months of 1989. Despite efforts by pro-incineration forces to get them to do so, the League of Women Voters refused to give up its right to participate in the DEC permit hearings, because it had problems with parts of an agreement the SWDA had drawn up with the Environmental Management Council (EMC) in February. In the same month, Canton Village trustees—pressured by the pro-incinerator County Planning Director Mary Verlaque,

20. Taped interview with Klaus Proemm, Canton, New York, December 11, 1990.

who also happened to be a village resident—nevertheless refused to rule on just how long Canton residents could keep anti-incinerator signs on their lawns.

More ominous was the endorsement by the St. Lawrence County Medical Society, in March, of "the delay of a trash burning facility until recycling is subjected to a more complete test" (Zissu, 1989). Robert E. Buhts, vice-chairman of the St. Lawrence SWDA, publicly challenged the Medical Society and insisted that their members were "misinformed if they believe recycling isn't a main concern with the authority":

> We are giving recycling a chance and we are deeply committed to recycling. Recycling is a vital part of any system. On February 15, I held a special public meeting in Potsdam to explain about the incinerator and I didn't see Dr. [Brian] McMurray [spokesperson for Medical Society] there. He is not in the right frame of mind. His way of thinking is like two or three years ago. It's a shame. This kind of rhetoric is not doing any good to anybody. If they have concerns about health issues and have proof to back it up, I'd love to hear it. (Raymo, 1989)

To raise money and mobilize citizen opposition to the proposed incinerator, WOW activists organized presentations, potluck dinners, and a variety of other activities throughout 1989. In January, for example, WOW hosted a talk by Tom Webster, an environmental scientist for the Center for Biology of Natural Systems, on the incinerator's health-risk assessment (HRA) document. In April, it was a potluck dinner, after which Lee Wasserman, executive director of the Environmental Planning Lobby of New York, talked about new legislation involving alternatives to incineration being introduced at the time in the state legislature. Later the same month, a letter went out to all pastors in St. Lawrence County over the signatures of "Ellen and Paul Connett," which read, in part:

> We are writing to you, and other pastors in St. Lawrence County, to request that you inform your congregation about a very important rally that will take place on the lawn outside the County Court House in Canton on Monday, May 8th, at 6:45 P.M. We hope also that you will either read this letter of explanation during your church service or include it in your weekly hand-out.
>
> This rally is being organized by Work on Waste for the purpose of

> letting our legislators know how many people oppose the building of a trash incinerator in our County. This may be the last chance we'll get before the legislators are asked to underwrite the bonds for this incredibly expensive project.
>
> Work on Waste has been involved in this issue for over four years and the reason we have continued for so long, and against such odds, is our belief that we have the moral imperative to protect our environment from further abuse. . . . We believe our task is not to find a new place to put the waste but to find ways to stop making it.
>
> We believe that we are getting warning signals at both the global level and the local level that we can't run a throw-away society on a finite planet. Incineration merely burns the evidence. . . .
>
> We are addressing you because we believe that ethics is at the very heart of this issue. As Sister Ruth de Platney so eloquently put it at a 1986 hearing in Ogdensburg, "We have to stop living as if we were the last generation on earth."[21]

Using a wide variety of means, WOW activists spread word of this May 8 rally on the courthouse lawn in Canton against the proposed Ogdensburg incinerator. One of the most dynamic and widely respected WOW activists—Rachel Grant, a young Canton schoolteacher—had the following letter to the editor published in a regional weekly, the *County Courier*, the week before the rally and only a few months before her own death from cancer:

> If you're like I was for quite a while you're probably saying, "I really don't want the garbage incinerator, but I'm going to let the WOW people take care of stopping it. I don't have the time to get involved." Or maybe you're saying, "I would do something to stop the incinerator if I knew what would help, but it's probably too late anyway."
>
> Fortunately, it's not too late; there is something easy every one of us can do. On Monday night, May 8, at 6:45 all you have to do is join hundreds of your neighbors on the lawn of the Court House, Court St., Canton. You'll be giving about an hour of your time to show your county legislators that the majority of people oppose an

21. This WOW letter was included in a packet of the group's materials organized and kindly provided by Ruth Beebe for this project.

expensive, polluting incinerator. Not a lot of effort for clean air and saved money.

And in case you're thinking of waiting for the next time, don't. This may be our last chance. The county legislature will be voting on the bond money for the construction of the incinerator this summer.

The legislators have the final say on the incinerator because they control the money. And it's those controllers of our money we'll be meeting with on Monday night. Our plans are for the legislators to come outside on the lawn to see the people and hear what the people have to say.

If every person who opposes the incinerator comes out on Monday night, we'll be able to stop those incinerator plans. Our power will be judged by the number of people we bring out. If you stay home, you're counted as a supporter of this expensive, polluting incinerator. If you come out, you're counted as supporter of the cheaper, cleaner solution—recycling, composting and reduction of our waste. Your vote is counted by where you are Monday night. And if there are enough of us on the Court House lawn, we can stop the incinerator.

WOW's activists were successful in getting a large turnout despite cold temperatures, a chilling wind, and overcast skies on May 8, 1989, as the following newspaper account of the rally explains:

> They came from all over, converging on the lawn across from the county's courthouse Monday night to urge the legislature to dump the proposed incinerator. . . . Close to a thousand people gathered to let members of the county legislature know they'd remember next November whether their pleas were heard.
>
> The great majority weren't the college students their critics had predicted would show up. Some brought their children by the hand. Others carried them in backpacks. A few reminisced about campus protests in the sixties. They carried signs and placards urging the county to drop its plan to build a $22 million incinerator. Only a few were curiosity seekers. . . .
>
> They were Work on Waste's constituency, citizens who said they wanted the legislature to know they cared enough to come out and show their viewpoint. They also wanted legislators to know that they

> were counting the legislators' votes, seeing who they'll try to unseat next November. . . .
>
> The legislators, eight of whom voted not to come out to listen to the crowd, were inside the stone courthouse for their monthly meeting. But those who did come out sat in folding chairs across the street from the crowd. They heard a procession of speakers urge them to recycle the county's trash.
>
> Brian McMurray, a Canton physician, rapped both the county and the *Ogdensburg Journal*, saying the press has "failed" to promote a dialogue on the issue. "WOW should be congratulated for their effort," he said. "I'm proud of their concern and their courage."
>
> Canton Mayor Marilyn Mintener, who said she was speaking as a private citizen, called the legislators who refused to come out to listen to the crowd "a disgrace." "I hope the voters will remember you in November," she said. . . .
>
> Chet Bisnett of Potsdam's Waste Stream Management said the county's private haulers are already promoting recycling and reuse of materials. . . . He rapped county officials for their flow control law which he claims makes it illegal to recycle some goods. (Reagen, 1989)

These pressures from the grassroots and private citizens were combining at the time with an increasing emphasis by the DEC commissioner's office on serious prior recycling before granting incinerator permits. Richard Grover told us: "Although there was no recycling competence in the DEC at the time this incinerator proposal was laid on the table in the early 1980s, that agency did start playing catch-up ball in this respect by the late 1980s." Pushed from below by pressures from WOW and from above by an increasingly pro-recycling ideology from the state, St. Lawrence County legislators and SWDA officials were forced to rethink their previous commitment to the incinerator. In May 1989, a surprisingly close vote on a landfill option in Lisbon, a community near Ogdensburg, revealed a rift in the SWDA on the incinerator option and encouraged opponents:

> In what may be evidence of a growing division, county waste disposal officials narrowly rejected a chance Tuesday to steer away from the controversial incinerator. In a dramatic 5–4 vote, the Solid Waste Disposal Authority defeated a motion to "explore" conversion of the

> planned Lisbon ash landfill into an all-garbage facility, thus dumping the incinerator. . . .
>
> Robert E. Buhts, who made the motion, said recent remarks by . . . Commissioner Thomas C. Jorling indicated the authority should consider "landfilling and aggressive recycling" as an option. . . . The DEC is warning SWDA that data supporting an all-garbage incinerator could be grounds for denial of the authority's permit. In other words, he said, the authority must look at the option, or risk being found at fault for disregarding the idea.
>
> While most members agreed that could be true, opponents said passage of the resolution would also send a message that SWDA is considering a break from the 10-year plan to build an incinerator.
>
> However, in rejecting the resolution by only one vote, [SWDA] members may have inadvertently sent out another, perhaps more important, message: there is a division among board members over incineration and the authority's current course.
>
> Such wavering on the centerpiece of the authority's plan is expected to further fuel efforts by such opponents as Work on Waste, which hopes to stop the incinerator by electing its own county legislators during the fall elections. . . .
>
> Mr. Proemm, an outspoken WOW member who was visibly pleased by the action, pointed out SWDA may have passed up a golden opportunity. "The DEC commissioner [by pointing out that SWDA should explore landfilling] may be warning the authority to make its case or risk not getting their permit," he explained. (Cummings, 1989b)

In an interview with another reporter, the same Klaus Proemm "said he was pleased the state Department of Environmental Conservation appears to be shifting its gears."

> Quoting a document distributed by the DEC, Mr. Proemm said incineration has been placed at the bottom of the agency's priority list, along with landfilling. "Prior to the memo, it had been somewhere in the middle," he said. "Things are changing all over," Mr. Proemm said. "The DEC itself is leaning further and further from incineration." (Lovett, 1989)

Opponents of the incinerator received a major assist in June 1989 when evidence surfaced that the figures in the health-risk assessment were modified in response to suggestions, or instructions, from SWDA Attorney Richard Cogen. Cogen's marginal notes instructing, along with other changes, that risk numbers be reduced to make it a more "positive" and "defensible" document were released to the press, causing incinerator proponents considerable political and legal difficulties.

> Numbers contained in the final health risk assessment for the proposed incinerator were lowered—in one instance by a factor of almost 10—in response to suggestions from a St. Lawrence County Solid Waste Disposal Authority attorney. . . . For example, the cancer risk to infants was reduced from 115 per one million to 14.6 per one million. . . . Through handwritten notes in the margins of the draft, Mr. Cogen told Environmental Risk Limited, a Connecticut company writing the assessment, to reduce certain figures, avoid other issues, and to reword sections. . . .
>
> But SWDA Executive Director Paul J. O'Connor said the attorney's involvement may not be significant in the overall context of the massive report. "I don't see any problem with what the attorney did, or the changes that resulted. Most of the changes are in response to points raised by the Department of Health, which he pointed out to the consultants," he said. Mr. O'Connor added the state Department of Health was concerned that risk figures in the initial version of the report were too high, due to faulty methodology and recently adopted emission standards which were not taken into account. . . .
>
> [O'Connor said:] "WOW's concerns about the attorney's involvement in the report are inappropriate. It's just an attempt to revert attention from their loss of credibility in assessing the health impact." . . .
>
> Dr. Paul Connett, leader of WOW, disagreed, saying the document is supposed to be an independent study of the risk associated with the incinerator. "Instead," he said, "SWDA got what it wanted: an advocacy document proving there is little risk to residents from the plant." (Cummings, 1989c)

Tom Plastino suggested that such incriminating information was due to political infighting within SWDA itself. "Doubts within the Authority and its staff produced these leaks [that made] SWDA look bad."

WOW's Connett demanded a full inquiry into the events surrounding the writing of the health-risk assessment and a probe into how the final numbers were devised. DEC Law Judge Daniel Louis made an unexpected exception both in allowing Connett to speak at "an unscheduled forum . . . during the issue conference on the proposed [Lisbon] ash landfill at city hall" and in assuring him that "he would take the matter under advisement":

> "I was not told to do this," [Judge Louis] said of allowing public discussion on the draft. "But since the commissioner [Jorling] is reviewing this, I will put these comments on the record for his, or anyone else's, use." (Cummings, 1989d)

Formally excluded from the permit hearings by Judge Louis, WOW organized a call-in and write-in campaign to Governor Cuomo during early September 1989, protesting its exclusion. A "WOW Emergency Appeal" was sent out to supporters, reading, in part:

> WE URGENTLY NEED YOUR HELP!
>
> The DEC permit hearings for the incinerator are set to begin on Oct 2, but WOW (the only group representing County residents on this issue) won't be represented unless we act quickly.
>
> We believe that WOW has been engineered out of the hearings for political reasons rather than the merits of our case. . . . We have to bring Governor Cuomo into the picture. We hope that he will insist that [DEC Commissioner] Jorling settle this matter objectively and fairly.
>
> YOU CAN HELP IN THREE WAYS:
>
> 1) WRITE a letter to Governor Cuomo indicating your disgust with the way that WOW is being excluded from the permit hearings. The arguments are listed in the attached letter. . . .
> 2) PHONE Governor Cuomo's office at 518-474-1288, try to speak with Governor Cuomo directly if you can, if not leave a message indicating your disappointment at the way WOW is being treated by the DEC. . . .
> 3) MAKE A GENEROUS CONTRIBUTION TO WOW. . . .

On the bottom of the page, volunteers were asked to check one or more of eleven categories including mailings, photocopying, door-to-door canvassing, babysitting, fund-raising, organizing a coffee meeting to show WOW's video, strategy, artwork, and related tasks. The letter attached, which WOW offered as a model, provided specific details criticizing SWDA's failure to plan for recycling "40% of the county's waste, as required by NY State law" as well as the methodology used in the health-risk assessment.

The permit hearings—the culmination of a ten-year process geared toward the construction of the incinerator—were held in early October 1989. At one point, SWDA Attorney Richard Cogen attempted to discredit Richard Grover, an expert witness for Waste Stream Management who argued that the SWDA's flow control laws prevented recycling efforts. Cogen repeated the story of Grover's forced resignation from his county job in 1979:

> Although the history of Mr. Grover's resignation from the county was new to Judge Louis, others attending the hearing expressed shock that SWDA raised the issue. "That's just plain dirty," said Waste Stream co-owner Chester W. Bisnett. "What happened 10 years ago is irrelevant." . . .
>
> As Mr. Grover's testimony continued for the rest of the day, Mr. Cogen repeatedly attacked his statements in an effort to portray Waste Stream's argument that recycling could already be under way as nothing more than speculation. . . .
>
> The main point, Mr. Grover emphasized, is that SWDA's flow control is preventing recycling efforts. . . . Mr. Grover said SWDA's consistent denial of requests from Waste Stream Management to remove certain items from the waste stream has damaged recycling efforts. (Cummings, 1989e)

WOW considered its grassroots political efforts rewarded when ten new members were elected to the twenty-two-person legislature during November 1989, including a number of candidates either explicitly endorsed by WOW or with positions that Grover characterized, in the interview cited earlier, as "favoring recycling and against importing trash—on the periphery of the issue without coming right out and saying 'I'm opposed to the incinerator.' " This was the legislature that would ultimately decide the incinerator's fate:

> The new board, which will include several newcomers who have expressed serious reservations about the proposed incinerator, will be asked to select a member to the Solid Waste Disposal Authority at their January meeting. . . .
>
> Sources say the next appointee to SWDA could play a major role in determining the future of the county's solid waste management system since the board is nearly evenly divided among incinerator proponents and opponents. (Martin, 1989)

It is instructive to consider these November 1989 election results more carefully, because WOW had not actively supported the majority of the newly elected legislators. Paul Connett told us:

> Of the ten new legislators voted into office in November of '89, we were able to get three people elected who were against the incinerator. There are twenty-two legislators, total. We always had hope, but it was a longshot because if you'd had a vote before this election [the incinerator] would've won 20–2, and we'd have been holding our breath even on the two. But after the election, two things happened: we got at least three people in there who were anti-incinerator and who were very bright and worked hard. The other thing, however, was that a lot of new people were elected who were not beholden to past commitments and who were open-minded. . . . They were elected on many different issues, but they looked at the incinerator from new perspectives. And that's where Bob Penski, a businessman, was so important. . . .
>
> By and large, most people in this county and in most counties do not believe they are going to be impacted environmentally by the incinerator. They think, "It's those poor suckers in Ogdensburg who are going to get it. . . . If most people in Ogdensburg seem to want it because the *Ogdensburg Journal* told them they wanted it, then what's the big issue? . . . Ogdensburg's seventeen miles away from us in Canton, and twenty-seven miles away from Potsdam—university towns from which the real concern against the proposed Ogdensburg incinerator was coming. . . . What got this county's twenty-two legislators was the enormous economic liabilities involved. . . . At the eleventh hour, Bob Penski, a local businessman who had not been involved in the incinerator battle and who was well respected for his economic abilities, decided to look at this from an economic perspec-

> tive and he realized this was an incredible ripoff. . . . He was an unpaid consultant serving as an independent voice who had a major impact on some legislators.[22]

In addition to their involvement in the county election, WOW activists also worked to bring international pressures against the proposed incinerator. One such project encouraged supporters to send out postcards to hundreds of sympathizers, encouraging them to sign "The Citizens Treaty Against Trans-Boundary Pollution" at noon, November 19, 1989, on the Ogdensburg Bridge to Canada:

> The Citizens Treaty will be a first step to combat the air, water and soil pollutants that cross the St. Lawrence River. The bureaucracies of both the American and Canadian governments abandon responsibility for pollutants once they cross political boundaries. PEOPLE OPPOSED TO WASTE INCINERATION (P.O.W.I.) of Ontario, Canada, and WORK ON WASTE will join forces under the Treaty to stop the hazardous waste incinerator proposed for Maitland, Ontario (just two miles from our border) and the garbage incinerator proposed for Ogdensburg.
>
> WOW invites you to be a signatory to the Citizens Treaty. Please come—with children and friends—to meet our Canadian neighbors. An informational and social gathering will follow the signing on the bridge.

Such projects, while intended to increase outside pressures on county decision-makers, were sometimes perceived differently from the way WOW activists intended. Tom Plastino, who had no WOW affiliation, commented:

> The "hands across the border" stuff was almost completely irrelevant to the debate here. If anything, like so many of the stunts WOW tried, it probably lost them more support than it gained. Incinerator proponents regularly played on the county's characteristic nativism by highlighting how outsiders (British professors, Canadians, the DEC, etc.) were trying to tell St. Lawrence County people what to do.[23]

22. Connett interview.
23. Plastino letter to Walsh (see note 16).

The following spring, in March 1990, a "Don't Burn New York" rally was sponsored by the New York Public Interest Research Group and Work on Waste in Albany. Approximately 500 people marched to the governor's mansion, demanding that state leaders stop garbage burning and supply more funds for recycling. Governor Cuomo spoke with some of the activists, insisting that the state could not stop incineration and landfilling until recycling was working much more efficiently. The activists countered that the state was not spending enough to make recycling work.

WOW's spring newsletter made the group's 1990 priorities clear: "We're planning lots of activities, all directed to the ultimate goal of an intensive grassroots lobbying campaign to convince the County legislators to vote NO on the incinerator." Such efforts, in conjunction with the new blood in the county legislature and the emphasis on recycling by the DEC, resulted in increasingly serious problems for project supporters. In May, for example, the St. Lawrence County Board of Legislators demanded final say over whether trash from other counties was to be burned at the proposed incinerator, but SWDA board members opposed handing over such power, leaving the issue unresolved (Tooley, 1990). The next month, DEC Commissioner Jorling ordered the SWDA to modify its flow control law giving it jurisdiction over the waste stream, presumably because of the SWDA's "failure to start up its recycling programs fast enough."

> Jorling's decision also suggests the authority consider building a smaller incinerator plant or expand its wasteshed by including a neighboring community or county. The idea of importing trash from a neighboring county would generate stiff opposition. Downsizing the incinerator could mean scrapping the entire project, breaking contracts with the current builder, increasing the size of the proposed landfill in Lisbon, and starting over from scratch. (Reagen, 1990)

The SWDA had frequently insisted that it needed the flow control law, which gave it total command of the waste stream, in order to make its bonds saleable (Cartier, 1990a).

As the July deadline for the county legislature's vote on the project's bonding approached, criticism against the incinerator continued to mount. Some legislators complained they did not have enough information to vote earlier, as the SWDA was urging, and legislature chairperson Betty Bradley commented in mid-June: "After what's happened in the past month, I would vote No" (Martin, 1990). The SWDA's executive director warned, however, that the authority faced "penalties between $500,000 and $3.5 million if it does not proceed with construction of its proposed solid waste incinerator by July 19" (Donnelly, 1990a). A WOW flyer mailed to "1500 supporters" in June, on the other hand, included the following itemization of what it labeled "SWDA's own estimates" for the incinerator system:

Upcoming bond issue	49 million
Bonds already spent	9 million
Grant from NY State DEC (your tax dollars)	6 million
Estimated interest over 20 years	90 million
Incinerator operating costs over 20 years	40 million
Total	$194 million

Then, in bold letters, the conclusion: "$194 Million Dollars—Minimum Estimated Cost of Incineration." The same WOW flyer insisted that citizen input was critical in pressuring legislators. Specifically, it advised:

> • Contact your county legislator. Tell him or her you're against the incinerator. Your legislator's name is printed on your mailing label. . . .
> • Come to the County Legislature meeting Monday, June 11, 7 PM at the Court House in Canton. This is their last meeting before the legislators will be asked to vote on the bonds at an as yet to be announced special meeting. WOW has invited a city councilman from Rutland, Vermont, to speak about the failed Rutland incinerator. Please come to show your support. . . .
>
> With your help we have a good chance to defeat the incinerator to make way for a sound system of solid waste management.

The League of Women Voters publicly expressed satisfaction with Jorling's emphasis on serious recycling as a precondition for an incinerator permit and with his criticisms of the SWDA. The statement continued:

> The conditions set forth . . . are quite similar to conclusions reached by the League and communicated to Administrative Law Judge Daniel Louis. . . . The League sees the decision as a positive action which will allow St. Lawrence County to draw all factions together and

> find a satisfactory solution to the solid waste problem. It is time to put differences aside. (Published in *The Plaindealer*, June 13, 1990)

The town of Lisbon, a few miles from Ogdensburg, also told the county legislature in mid-June that they did not want trash imported in order to make the incinerator financially feasible. A local citizen and the organizer of "Farmers Against the Incinerator," Don Hassig, said that Lisbon "would feel the brunt of the negative impacts that the garbage burner would cause."

> Hassig told the [Lisbon] board that he had spoken to county planner Mary Verlaque and was told that the county wouldn't import garbage from outside the county. But Hassig said he didn't buy this. He said that the three recommendations by New York State DEC Commissioner Thomas Jorling spell out that the county would have to either reduce the size of the project, abandon it or import garbage. Hassig says he believes the county will choose importation.
>
> "It doesn't look like they're going to do anything but import. Once they get a bill, there's no way they're not going to import. They almost have to import or they won't be able to sell the bonds," said Hassig. (Purser, 1990a)

For members of the county legislature, as the day of decision approached, economic considerations became increasingly salient. Legislator Clyde L. Morse, for example, told a town board in mid-June: "The financial burdens of the proposed incinerator project are beginning to outweigh all other issues that have been attached to the garbage burning facility. . . . The economics is getting real questionable" (Purser, 1990b).

The county's Environmental Management Council (EMC), which had supported the project with certain conditions back in 1986, voted unanimously in late June 1990 "to reconsider its position on the proposed incinerator, ash/bypass landfill at a committee meeting on July 5."

> EMC Chairman Ed Fuhr supported Plastino's idea of the board rethinking its position and coming up with a firm, up-to-date position before the County Legislature is asked to vote on the $49 million bonding issue on July 19.
>
> The council passed a motion to have the solid waste committee and any other members attend the meeting to look at the past deci-

> sion on the incinerator, what changes have been made since 1986, and decide what position the EMC would take now. All members attending the committee meeting will have full voting power. (Cartier, 1990b)

Despite such negatives, a poll of county legislators by Park Newspapers in late June concluded that the votes necessary for financing the proposed incinerator were quite secure:

> A survey of county legislators shows that while some say they are undecided on how they will vote, the county's solution to its garbage problem still has enough backing from legislators to give it the 12 votes it needs for passage.
>
> The Park Newspaper poll shows Carl Ashley (R-Ogdensburg), Duana Carey (R-Stockholm), Allen Dunham (R-Hammond), William Lacy (R-Gouverneur), Thomas Luckie (D-Ogdensburg), Stanley Morrill (D-Hermon), Charles Romigh (R-Massena), James T. Smith (D-Canton), Robert Snider (R-Star Lake) and Stephen Teele (R-Lisbon) all plan to vote in favor of the project. Chris Well (D-Ogdensburg) was unavailable, but when last contacted, he favored the project.
>
> Donald Smith (D-Louisville) said he was upset the newspaper was taking a poll on legislators. He said that he could be put down as undecided, but other legislators say Smith has been a strong supporter of the legislature backing the authority.
>
> With 22 legislators on the board, any measure passes with 12 votes. But supporters of the project are predicting that when the roll call is made on the project, the board will see more than 12 casting their vote in favor.
>
> Supporters say that some of the legislators who have said they are undecided will vote yes when the issue comes to the floor. (Cartier and Reagen, 1990)

Amid such upbeat speculation and encouraging polls on legislators' voting preferences, however, news that the consulting firm guiding St. Lawrence County's incineration plans had recently advised three Maryland counties to avoid trash burning raised additional problems for proponents. Virginia consultants Gershman, Brickner & Bratton (GBB), explained this discrepancy in recommendations by referring to "vast economic and other differences between the two regions," but project opponents used this information to support their arguments against incineration.

> GBB was initially hired by St. Lawrence County more than a decade ago after [the SWDA] was created and has guided the incineration project ever since. . . . Timothy Bratton, a partner in the firm, said Thursday that the consultants have recommended that Calvert, Clark and St. Mary's counties in Maryland avoid incineration, concentrate on recycling and consider building a "mixed waste-processing facility to produce a compost product." (Donnelly, 1990b)

Although the article went on to list differences between the two study areas addressed by GBB—including "weak energy arrangements in Maryland . . . [more] landfill capacity in the Maryland counties, . . . [and the fact that] Maryland counties are barred by 'anti-double-dipping' provisions of the U.S. Tax Code from issuing tax exempt bonds—the firm's reluctance to endorse incineration was viewed by many opponents as an indication that the GBB was speaking out of both sides of its mouth.

Despite the confident predictions by project proponents, the bonding vote by legislators on July 9 was 11–11, and when weighted on the basis of population represented by the legislators as legally required in the case of ties, the incinerator was defeated. Explanations for this surprising outcome were somewhat different. We turn, first, to interpretations by two project proponents before hearing from three who opposed it. County Planning Director Mary Verlaque explained: "Three days before the vote, the Environmental Management Council said they were concerned about the economics of this project, and that didn't help with the legislators' votes." She said that of the ten new legislators taking office in January 1990, "seven or so voted against the project. . . . The anti-incineration hype was really the cause of its defeat." A leading supporter of the project, Verlaque explained that the challengers' focal issues shifted from health to economics but that she was shocked to see how much impact this grassroots criticism had on the newly elected legislators: "Once opponents started to instill doubts, the public kept slipping away, but nobody expected the legislators to take the protestors so seriously. . . . Even though WOW was based in Canton, they went door-to-door throughout the county fighting this thing."

Another project proponent, however, disagreed with Verlaque's attributing so much influence to WOW in the defeat. Bill Sutkus, long involved with this project through his work on both the Environmental Management

Council and the SWDA, disagreed with any conclusion that "the defeat of the incinerator was due to WOW's influence on the '89 election and the July '90 vote." He elaborated:

> At the least, it would be necessary to analyze the '89 election district by district, and the July '90 vote legislator by legislator. My guess is that in the election, three of the winners met with WOW approval and probably only one won due at all to WOW's work. In the July '90 vote, perhaps three or four legislators voting "No" would have given part of the credit to WOW. On the other hand, a couple of the legislators one would have expected to vote "No" would have credited WOW with their enthusiastic "Yes" vote. It is not irrelevant to ask if the vote against the incinerator would not have been greater without WOW.[24]

Tom Plastino, also a member of the Environmental Management Council, disagreed with Sutkus's suggestion that the incinerator would have been defeated without WOW's work, although he was not himself a WOW supporter. Plastino offered a detailed analysis of the legislators' final vote:

> Without WOW's work, there's no doubt in my mind that the incinerator project would have been successful, and Bill Sutkus is not correct about there being more votes against the incinerator without the Connetts' and WOW's work. . . . But WOW was only one of several factors contributing to the incinerator's demise. WOW certainly did keep the pot boiling for three or more years, but for most of the legislators it was only one contributing element. . . . WOW's efforts were only a part of what it took to blow the project out of the water.
>
> It took 11 votes to beat [the incinerator], and with weighted voting, they had to be the *right* votes. I can only attribute *at most* six nay votes to WOW influence. . . . Very few of the legislators who voted Nay were swayed by WOW's environmental arguments. . . . At least six were swayed by severe doubts about the project's economics. . . . Bob Penski's influence was the new element on this score in the spring of 1990. His fresh voice, together with the declining credibility of GBB, when added to the steady undercurrent of eco-

24. Written comments by Bill Sutkus on an earlier draft of this chapter, November 3, 1993.

> nomic doubts, tipped at least four legislators against the project (Bradley, McFaddin, Paquin, and MacLennan). . . .
>
> Legislators like Purvis, Moore, Darmody, and Pierce, I think, were more swayed by the importation red herring which is really just nativism dressed up as environmentalism, . . . [and] they didn't need WOW to convince them of that. . . .
>
> Teele had to vote Nay if he was to retain his seat—that wasn't WOW's doing. The biggest mistake the incinerator proponents made was for the Republicans not to pressure Teele to vote Yes. They thought they had the votes to pass it without him, so they allowed him to vote Nay so he could survive in Lisbon. . . . Although it's invidious to single out any one vote as most important when you need every single one, I still think that the persons most responsible for the defeat of the incinerator were the Republican leaders, because they failed to bring the necessary pressure to bear on Teele.
>
> That leaves Morse and Tomlinson as being the people who were probably influenced by WOW. . . .
>
> One last point here: WOW *did* talk economics, but their economic data had about as much credibility as that adduced by SWDA and its hirelings. The real economic doubts were planted by people like Penski and some of the EMC board members—people who had either not engaged in the debate publicly before or who had always labored to give the impression of being objective.[25]

Richard Grover's interpretation of the election and legislators' vote presents a somewhat different perspective. Reflecting on this outcome, and emphasizing the distinction between the communities most opposed to the project and the proposed host community, he explained:

> Remember that what started out as a voice in the wilderness has snowballed into a massive community outcry, centered in Canton and Potsdam—two college communities in this county.
>
> Where the incinerator is going to be built, in Ogdensburg, people are pretty much silent, and it was largely viewed as an "intellectual issue" in the county. . . . Some would say "Look at the people

25. Tom Plastino, excerpts from his written comments to the authors and from a follow-up telephone conversation (see note 16). The telephone conversation focused upon the written comments, seeking more details and further elaboration.

> screaming about this issue, look at where they live and what they look like—they've got beards. . . ."
>
> In the fall of '89 there was a major turnover in the St. Lawrence County legislature, and the incinerator was a key issue. . . . We have a twenty-two-person legislature, with districts set up on the 1970s census. . . . This county has a weighted voting formula to adjust for 1980 census changes in population here, which only came into effect in the event of a tie vote. . . . There were seven or eight candidates that ran on the incinerator issue, and the ones that won upset long-time legislators. . . .
>
> Schoolteachers and carpenters were elected who favored serious recycling before deciding on the size of an incinerator and who opposed importing garbage. . . . So when it came to the vote, they opposed it because they knew the county would have to import to get enough trash to make the incinerator economical. . . . They had to vote NO. . . . That's where the 11–11 vote came, and with the weighting, the project was killed.

Paul Connett's own interpretation of the surprising vote by the county legislature emphasized the gradual erosion of support for the project during the final week.

> Surveys by the radio stations and newspapers were coming out before the incinerator vote for bonds—after they had already gotten their permits from the DEC—and this is where the economic liabilities were key. The week before, it was 12–6 in favor of issuing the bonds, with four undecided. A couple of days later, it was 12–8. The day before the vote, it was 12–10, so we were getting the uncommitted against the incinerator as we progressed, but we couldn't dislodge any of these twelve votes for the incinerator bonds. The day of the vote, one guy switched—so when it came to the vote, it was 11–11. . . . [The one who switched] was not in favor of the incinerator, but he had been worried that if he didn't vote for the incinerator the DEC would close down our landfill—after he called the DEC and learned that wasn't true, he voted against.

In addition to their work with the local Work on Waste, the Connetts were also now editing a national anti-incineration newsletter, "Waste Not." In its first postvote publication, July 12, 1990, they offered a capsule summary of their perspectives on the struggle, under the headline "The Defeat of the Incinerator in St. Lawrence County":

> The Editors of this newsletter have had more than a passing interest in the St. Lawrence County incinerator battle! It was this battle, in our own county, that first got us involved five and a half years ago with the waste issue. Indeed, it was this battle which led to our reaching out to other groups around the country involved in similar battles. In turn, this led to the formation of the National Campaign Against Mass-Burn Incineration and for Safe Alternatives in 1985 by groups in New Haven (CT), Holyoke (MA), Rutland (VT), Claremont (NH), and St. Lawrence County (NY), and to the series of newsletters produced in 1986–87 (does anyone remember the 200-page newsletter!), which evolved into *Work on Waste—USA*, the newsletter *Waste Not*, [Connett's] research with Tom Webster on the build-up of dioxins in the food chain, the series of video tapes co-produced with Roger Bailey, and countless speaking engagements.
>
> Waking up last Tuesday morning, the day after the Legislature voted the project down, was like waking up in a new world. All of a sudden one could contemplate using one's energy on positive things—like helping to work on a county-wide waste management plan we could be proud of, instead of the energy draining exercise of trying to stop a moving train greased by short-sighted officials and driven by out-of-state consultants.

The same July newsletter summarized what WOW had learned from the struggle. Its editors endorsed the advice of Lois Gibbs and her co-workers at the CCHW:

> You don't win these battles with experts, with lawyers, or with the regulatory process; you win them by working together to educate the community on the issue and by using the political process. Local politicians must be held accountable for local planning decisions that impact on the community—they must not be allowed to hide behind unelected "authorities" or faceless bureaucrats. ("Waste Not," July 12, 1990)

The editors also addressed the issue of the relationship between science and politics in such disputes:

> In matters of environmental regulation, science is good for one thing only: to convince your key allies in the political fight that you are on solid ground. If that sounds defeatist coming from a scientist, it is not. While we are appalled that the key regulatory agencies which are supposed to be protecting our health and environment are more beholden to political and industrial pressures than they are to science, we will continue to work to bring science back into government. Meanwhile, we have to stress the importance of working together to educate our communities and to work politically at the grass-roots level until genuine leadership emerges with backbone enough to stand up to polluting corporations and the multi-billion dollar waste industry.

Attributing the project's defeat solely to the work of any single individual or group would be caricaturing an obviously complex process, according to County Planning Office staffer, Keith Zimmerman, who described himself with excessive modesty as "an individual somewhat knowledgeable about the history of the incinerator issue in St. Lawrence County after 1987":

> There were many factors contributing to the defeat of the facility, and I don't believe that, in this instance, it was solely the efforts of WOW or other anti-incinerator campaigners that "won the day." . . . Critical issues were the economics, the eroding credibility of [the SWDA's] consultants, the ongoing possibility that the County might import trash, etc. . . . While these were all issues identified by WOW, other groups were also questioning certain aspects of the proposal, and in my opinion they carried more general credibility to the discussion than did WOW.[26]

Most informed observers, including Zimmerman, generally agreed with Plastino's assessment that "without WOW's work . . . the incinerator project would have been successful." If the legislators' votes were the final blow, earlier factors—such as Waste Stream Management's lawsuit delaying the proposed siting for years, the election in 1989 of new legislators, changing economic realities, political pressures by other groups and individuals in the county, and the shifting policies of the state's DEC—have also been men-

26. Keith Zimmerman, letter to the authors responding to an early draft of this chapter, November 19, 1993.

tioned as major contributors to this defeat. While any simple attribution of the outcome to WOW's work is suspect, so also is the suggestion that this project would have been defeated without that work. This social movement organization's grassroots opposition activity prepared the public soil, so to speak, to support various challenges to the incinerator project, whether or not they were directly connected with WOW members or supporters. It also was the organizational carrier of such challenges after they surfaced—the county's collective memory, so to speak, for its anti-incineration case. If WOW sometimes served merely as an organizational cheerleader for important WSM lawsuits or economic challenges to the project by such individuals as Bob Penski, and even if WOW sometimes alienated potential allies with its strategies and tactics, few informed observers dispute the assertion that without WOW the incinerator would have been built. WOW coordinated the countywide opposition to the incinerator, which continued to increase during 1988 and 1989 as the St. Lawrence County Medical Society, local trash haulers, young mothers, farmers, the League of Women Voters, and other groups spoke out against it. The search for trash disposal alternatives in St. Lawrence County was back to square one after this July 1990 vote, and that signals the successful termination of this final grassroots challenge for our purposes.

The open-ended questionnaire responses from WOW activists revealed the wide variety of work done.[27] In addition to the obvious efforts of WOW activists—organizing, writing, researching, telephoning, lobbying legislators, fund-raising, public speaking, election work, demonstrating, etc.—some mentioned things like "working to establish recycling in my community," "taking notes at SWDA, DEC, and other such gatherings," and "organizing the Pete Seeger benefit." Important "turning points" and "key decisions" mentioned by a number of responding activists included litigation slowing the project, political work on elections resulting in a realignment of the legislature, and the decision to focus on economics more than on "complicated" environmental issues. One activist observed: "The

27. We received eleven questionnaires back from the twenty sent out, and in the process of trying to increase the response rate we were told that one of the St. Lawrence activists was discouraging others from returning theirs. One activist explained apologetically, "She's become paranoid from her involvement in all this." Although we are not aware of any published systematic discussion of this topic in the social movement literature, it is not uncommon for some grassroots activists to be uncooperative with academic researchers whose funding may be resented, even if it comes from what academics regard as "neutral" sources, or whose future "book" may be perceived as exploitive and resented.

lawsuits bought time and created frustration for county legislators and SWDA members, eventually causing a split in the legislature as well as in SWDA support for incineration." Klaus Proemm endorsed this notion and emphasized the critical nature of both the Waste Stream Management court case against SWDA, and another by WOW in helping slow the siting process enough to allow for the election of new legislators.[28] Another emphasized the effectiveness of economic arguments: "The overall cost to the county and to individuals was easier to convey than technical environmental concerns, and made a dramatic difference in public opinion." Still another mentioned "the large public meeting of over 1500 people on a cold, rainy day with the county legislature" as representing an important curve in the road to success. Two WOW activists also considered as critical the involvement of the medical profession in questioning the incinerator.

Again in St. Lawrence, what county spokespersons understood as a citizen advisory committee for the incinerator project was quite different from the ideal type described by industry consultants. In this case, officials considered their Environmental Management Council a CAC, but it did not play much of a role in the incinerator project except for its eleventh-hour 13–2 negative vote a few weeks before the legislators decided. Nine of the thirteen CAC responses we received mentioned this vote to endorse recycling and to oppose the incinerator project as the CAC's only critical decision. One labeled it "strong and overwhelming advice to our county legislature to turn down the incinerator."

The following excerpts from the numerous additional comments appended to these CAC questionnaires provide some useful insights regarding the evolution of perspectives among its members. Each excerpt is from a different respondent.

> I was originally in favor of the plant, but after a few months on the EMC the evidence changed my mind.
>
> We were rather accepting of the project *until* the incinerator people wanted to start *importing* garbage.
>
> I became convinced that the incinerator was not part of any larger system that included recycling, composting, and reduction. It was designed to be the only answer to our disposal problems.

28. Proemm interview.

The economics of the project did more to cause its failure than the environmental factors, but the unknown environmental impacts were enough to tip the scales.

Overview and Continuities

Even though most of their physical and structural characteristics in Table 7 might have predicted otherwise, both these attempts were defeated. The only indicator that suggested any problems for these particular sitings—that each left open the possibility that outside trash would be imported—did, in fact, become an important issue in both cases. This is the first time we have encountered CACs associated with a defeat, and it happened in both cases. But we have now said enough about confusion in the use of this CAC label to justify dismissing it as a critical variable influencing outcomes, except to note that when a number of individuals whom proponents consider CAC members openly oppose the project, as happened at both New York sites, that bodes ill for siting.

Our field data also help us understand better the striking framing differences on incinerator issues between these two particular backyard communities revealed in Chapter 3's telephone surveys. Now we realize that the citizens opposition group RECYCLE was centered in Broome's backyard community of Kirkwood, in contrast to its counterpart in St. Lawrence County's Ogdensburg, where there was no organized resistance because most residents supported the project. Thus, despite a remarkable similarity in demographic characteristics between the two communities, it is not too surprising to find the Kirkwood backyarders reporting higher levels of anti-incineration networking and framing than those in Ogdensburg.[29]

Effective protest mobilization in these counties actually emerged via quite different routes, even if both outcomes were similarly determined by surprising votes in their respective legislatures. In each, the national recycling debate—which became serious only in the mid-1980s—provided a master

29. As noted previously, these two New York projects were still in process when our surveys were conducted. In the northern St. Lawrence County along the Canadian border, the county legislature had voted negatively on the project a few months before, but some proponents were still talking about the possibility of modifying and reconsidering the project. In the southern part of the state, the Broome County telephone survey took place more than a year before the final decision on that project was made.

protest frame. In Broome, a county executive who spoke out against the incinerator before his election displaced a pro-incineration incumbent, and then as a newly elected executive became both a central player and a unifying symbol for three grassroots protest groups. The DEC commissioner's insistence that Broome downsize or import trash as a precondition for a permit helped turn all but the most committed legislators to alternative considerations, and yet the county executive's veto might still have been neutralized had just one legislator's request for more careful study been honored. Leadership of the St. Lawrence opposition was less mainstream, and various factors working in conjunction with WOW were identified as contributing to another shocking last-minute vote against the project in that county's legislature. The DEC rulings, which played into opponents' hands in both cases, were very important. Some observers even regarded them as critical.

Instead of the four sitings and four defeats originally envisioned, then, we are left with three sitings and five defeated projects. The possibility for such an outcome was built in to our research design from the beginning. One pro-incineration consultant even suggested that this result "probably says something about the recent national trend against incineration." As we have seen, however, a mere change in two legislators' votes could have left us with five sitings and three defeats.

What noteworthy resemblances do we find among these New York defeats and the previous ones in Cape May County (New Jersey), Lackawanna County (Pennsylvania), and Philadelphia? In each case, there was considerable organized resistance from outside the target area, and protests were accordingly framed to appeal to potential allies throughout the counties rather than being focused upon the backyard areas. In each of these five cases, there were also elite cleavages and an emphasis by challengers on political rather than legal strategies.

This chapter concludes our case study comparisons. Sometimes the details and idiosyncrasies of the different cases threatened to overwhelm us, but now it is time to ask broader questions covering all eight cases.

8

Implications: Theoretical and Practical

Can we identify critical factors distinguishing successful incinerator sitings from defeats of such projects by grassroots protests? Or should we have saved ourselves considerable trouble by simply accepting the observations of people on both sides of these controversies who emphasized the unique nature of each siting attempt and doubted that any valid generalizations about them could be made? Some of the bright undergrads and graduate students who read earlier versions of the previous four chapters said they felt confused, at times, with the large caste of characters and the variety of subplots involved in the eight case histories. Now that we have completed our examination of both the survey and fieldwork data, have any patterns emerged to help explain the different outcomes? In other words, why were incinerators successfully sited in Pennsylvania's Delaware, York, and Montgomery Counties but defeated in the five other cases (Pennsylvania's Lackawanna and Philadelphia Counties, New York's Broome and St. Lawrence Counties, and New Jersey's Cape May County)?[1]

1. The unexpected defeat of the Broome County project not only tilted our sample but also resulted in a noteworthy professional experience for us. In response to a request by the editor of *Social Problems* that we submit something for a special issue on environmental justice, we sent an article comparing a defeated project in Philadelphia with a

We respond to this central query by focusing upon three related questions. The first question returns to the theoretical debate about the relative importance of indigenous versus outside resources and the problems posed at the end of Chapter 3: How do our backyard survey data fit into the expanded perspectives provided by the detailed fieldwork in Chapters 4 through 7? Although Chapter 3's aggregate comparisons revealed that in backyards where projects were defeated the issues were generally framed more negatively, the survey data also raised more questions than it answered about why such outcomes emerged. In the light of what we learned from the fieldwork, the telephone survey results now make more sense.

The second question is related to the first: Which of the numerous differences were critical in accounting for outcomes? From the myriad of fascinating details associated with specific cases, can we identify any overall patterns? We shall draw upon both the fieldwork data and the telephone survey data to identify such commonalities.

The third question elaborates some theoretical consequences of our findings for environmental sociology and for contemporary social movement theory.

As a way of summarizing the practical consequences of our findings, we conclude by offering separate hypothetical memos to "proponents" and "opponents" of such projects.

Survey Results in Light of the Fieldwork Materials

The data from our fieldwork provided essential insights into the actual dynamics of each siting struggle initially considered in Chapter 3's telephone surveys, and they also made it easier to understand those survey results. The combination of survey and fieldwork data was particularly useful for illuminating the broader dynamics of each case. For instance, an item in the telephone survey asked backyarders whether they "belonged to any organizations or other groups that were against the trash plant" (see the Appendix for the precise wording of items). Because less than 10 percent of any popu-

---

"likely incinerator siting" in Broome County. After our submission was favorably reviewed and accepted, however, the Broome County project was defeated at the eleventh hour (Chapter 7). We made a last-minute change to the Montgomery County case as our example of a successful siting, before submitting the final version, published in February 1993.

lation typically become actively involved in social protest, we expected these results to be in the single digits. Contemporary social movement theory, as well as specific indicators in the Cerrell Report, also prompted us to expect more successful protest mobilization in backyards where such projects were defeated. While the aggregate data revealed a tendency in this predicted direction (3 percent in sited areas and 5 percent at defeated sites), it was not statistically significant—and, if one assumes that indigenous resources are critical in determining outcomes in such conflicts, remains puzzling.

If we examine responses to this item in light of what we now know from the fieldwork, however, the picture becomes clearer. The aggregate data was influenced by the fact that St. Lawrence County's successful opposition came entirely from outside the backyard area in Ogdensburg, which had *no* backyarders reporting that they belonged to a protest group. Findings like these show that backyard survey data is misleading when viewed in isolation from broader countywide processes identified through fieldwork.

Survey results from the successful sitings in Pennsylvania's Delaware, York, and Montgomery Counties also become more interpretable in the light of the fieldwork. Some of the questions raised in Chapter 3 about differences in our survey data among these sites can now be answered. Because the Delaware County project in Chester faced relatively little organized grassroots opposition, fewer respondents heard of any public meetings on the incinerator issue. An explanation for this absence of serious protest mobilization was that anti-incineration organizing was undermined by local authorities, who were considering an even larger incinerator of their own. In York, the fieldwork data also aided our understanding of why the backyarders' framing of issues tended to be more pro-incineration than at either of the other two sites where plants were built. We learned from the fieldwork that, especially during the first year or so of its struggle, ROBBI, the York protest group, adopted a vague and compromising position toward the proposed incinerator. In the third case, Montgomery County—where respondents were more educated and also more likely to have attended a public meeting on the incinerator issue—the fieldwork revealed that these backyarders found themselves relatively isolated in their county and unable to mobilize support from surrounding townships. Despite mounting the most serious backyard challenge of the three sited incinerators, these Montgomery County backyarders were also tardy in pressuring their local officials to oppose the project, which was championed by a politically powerful county commissioner.

Initially puzzling survey results in backyard communities where projects were defeated are also more understandable when illuminated by the fieldwork. The fieldwork's most remarkable contribution was the insights it provided into the dynamics of the St. Lawrence County struggle, where the projected host community, Ogdensburg, favored a trash plant that was ultimately defeated by organized outsiders from elsewhere in the county. In addition, the fieldwork data also helped explain why the percentages of minority residents in backyards within Cape May and Philadelphia were unimportant outcome predictors: resistance was quite broad-based, instead of being limited to these backyard communities.

Such insights from the fieldwork relativized the backyard survey data and showed how important it was to view the latter within a larger county context. For example, we considered one backyard community (Chester, in Delaware County), where local officials opposed a project from its beginning to no avail; another (Plymouth, in Montgomery County), where a grassroots protest group persuaded local officials to reverse their initial position and oppose a project, again with no success; and a third, where a backyard community was united in its willingness to host an incinerator (Ogdensburg, in St. Lawrence County), only to see that project defeated by organized opponents from other municipalities in the same county. Such outcomes accentuate the importance of considering the broader context within which indigenous resources, framing, and networks are situated. We do that next, in an effort to identify patterns across these different siting processes.

Critical Differences Between Sitings and Defeats

Which of the many variables we have considered, including both backyard areas and surrounding counties, were crucial in accounting for outcomes? We may start by asking whether timing was pivotal in explaining which projects were eventually completed and which were defeated. For example, in view of the historical evolution of opposition to WTE incinerators, we have noted that projects initiated before 1980 in the United States encountered virtually no organized grassroots opposition and also that relatively few have been successfully built since 1990—as a result of the national anti-incineration movement, which blossomed after the mid-1980s. But how about that focal decade of the 1980s? How much difference did timing make during the years when national momentum against incinerators was growing? Did the year the project was announced during this decade, the time opposition to it first emerged, or the span of time between the two account for the different outcomes? Figure 2 is an overview of the eight projects, indicating the times of the initial public notice, the time of the first challenge to the project, and the outcomes. The three successful sitings are listed first, in the order they appeared in the preceding chapters; the five defeated projects follow, in similar sequence.

If incinerator sitings were relatively routine before the 1980s and extremely difficult after 1990, Figure 2 suggests that the timing of project announcement or protest emergence during the 1980s was not critical. Factors other than timing were decisive for these cases during that decade. Observers emphasizing the central importance of early protest mobilization may call attention to Delaware County because it had its first organized challenge more than five years after that project was initially publicized, and even then the protest was in favor of Chester's own incinerator. And yet in the other two successful sitings, Montgomery and York, the amount of time between the first public notice of the project and organized protest was not obviously longer than in the cases of most of the defeated projects. On the other hand, one of the effective citizen challenges, St. Lawrence, had more than three years between the first public notice and the emergence of orga-

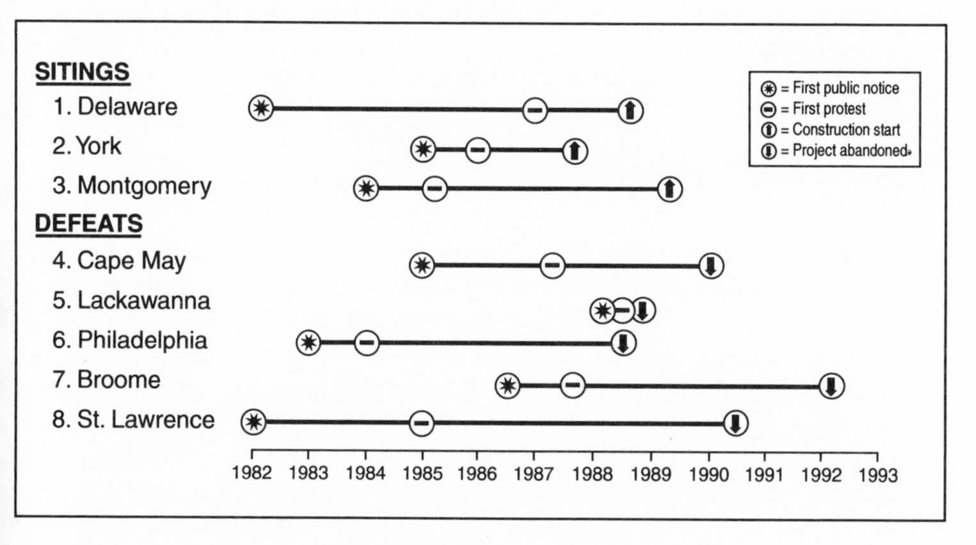

Figure 2. Timing and spacing of first public notice, organized protest, and outcomes.

nized opposition, and Cape May activists were also slow to mobilize. The quick demise of the Lackawanna County project is indeed noteworthy, and perhaps the outcome would have been different if the project had been begun a few years earlier. Among the early starters here, one first publicly discussed in 1982, Delaware County, was successfully sited, but another also initially publicized around the same time, St. Lawrence, was defeated. It is possible that a larger sample of incinerator projects might suggest a pattern, but these data do not reveal any relationship between timing and outcome.

Before turning to some of the other factors involved in such sitings, we can eliminate two seemingly influential ones. There was no systematic prior recycling at any of the sites so that it is not included on our list of "variables" (Table 8). Its absence was a "constant," even though recycling and its consequences were important considerations in most of our cases. And because only one project, Lackawanna, was not officially sponsored by its respective county, we cannot address the question of possible differences between county-sponsored and private attempts to site incinerators.

Distributions of selected variables across the eight sites, either emphasized by others or suggested by our research to be important in such processes, are listed in Table 8. Three variables (1–3) are from the backyard surveys, six (4–9) are the structural factors, or site characteristics, discussed at the beginning of Chapters 4–7, and seven (10–16) are from the fieldwork. While our primary goal here is to identify outcome predictors for these eight sites, we have no reason to doubt their relevance for similar projects across the United States.

Backyard Indicators

The telephone survey data from randomly selected backyarders within a few miles of each project led us to conclude that, despite some statistically significant differences between sitings and defeats, these data raised more questions about the outcomes than they answered. Because interested readers can refer to the more extensive data in Chapter 3 and in the Appendix, we limit our comparisons in Table 8 to three broad indicators of backyard sentiment and demographics.

The first two comparisons display the number of respondents in each backyard reporting themselves in favor of, or opposing the incinerator project (these numbers are also percentages, because the total number of respondents at each site was 100). While the percentage of those favoring the

Table 8 County comparisons on variables associated with siting outcomes

	Sited			Not Sited				
	Delaware	York	Montgomery	Cape May	Lackawanna	Philadelphia	Broome	St. Lawrence
A. Backyard indicators								
1. Favor project (%)	26	28	25	17	25	17	10	32
2. Oppose project (%)	17	11	26	37	24	41	64	22
3. Minority (%) (3 miles)	36	0	10	23	3	19	2	1
B. Site characteristics								
4. Size (tpd)	2,688	1,344	1,200	500	1,000	2,250	571	250
5. Import waste?	Y	Y	Y	?	Y	N	?	?
6. Two-mile population (thousands)	30	30	100	1	25	250	2	12
7. Annual compensation (millions)	2.0	0.3	0.35	2.0	2.0	0	0.25	0.075
8. Advisory committee	Y	N	N	N	N	N	Y	Y
9. Previous protest?	N	N	N	N	N	Y	N	N
C. Fieldwork Data								
10. Political champion?	Y	N	Y	N	N	Y	Y/N	N
11. SMO within one year?	N	Y	N	N	Y	Y	N	N
12. Years of protest	0	2	4	3	1	3	5	4
13. Outside assistance?	N	Y	Y	Y	Y	Y	Y	Y
14. Wide protest base?	N	N	N	Y	Y	Y	Y	Y
15. Primary SMO focus	NA	L	L	P	P	P	P	P
16. WTE split elites?	Y	N	N	Y	Y	Y	Y	Y

NOTES: SMO = Social movement organization
NA = Not Applicable
L = Legal
P = Political
N = No
Y = Yes

project hovers near 30 percent in the backyards where projects were sited, the fieldwork data explained why two defeated projects—Lackawanna and especially St. Lawrence—also had percentages of backyard respondents favoring the project similar to the three successful sitings. Conversely, the percentage of backyarders opposing these projects may also be a deceptive indicator of outcome—for example, there were more backyard opponents in Montgomery County where an incinerator was sited than in either St. Lawrence or Lackawanna, where projects were defeated. Such anomalies, as we have seen, are accounted for when we realize that the preferences of backyarders were seldom decisive factors influencing these outcomes.[2]

The third variable in Table 8 displays the percentage of backyard respondents belonging to any category other than "white." While the relatively large percentage of minority members at the sited Delaware County project may initially suggest support for an "environmental racism" argument, the fieldwork revealed that African American municipal officials there favored their own incinerator project over the county's plant. As we look across the row, we also see that the next highest percentages of minorities were in communities where projects were defeated: Cape May and Philadelphia. Such data do not support a straightforward variant of the environmental racism argument, that the higher the percentage of minority members in an area, the more likely a siting is to be successful. Yet neither do they completely undermine more nuanced statements of that general position, emphasizing the likelihood that such projects are more apt to be attempted in areas already zoned "industrial," where disproportionate percentages of poor minority group members live.

Site Characteristics

The next six variables in Table 8 are familiar from the previous four chapters. The three completed plants were among the top four in terms of size

2. With the exception of Broome, which stands out from the other sites on the first two variables in Table 8 because of its low percentage of respondents in favor and the remarkably high percentage of those opposing, more than 40 percent of the backyarders at each site did not report themselves as clearly favoring or opposing the incinerator project; such people said they had mixed feelings, no opinion, or were unaware of the conflict taking place in their own backyards (Chapter 3, note 1). In the wake of the Three Mile Island accident and evacuation, however, only 5 percent of the randomly selected respondents reported a similar "undecided" position on nuclear power issues (Walsh and Warland, 1983), which suggests a need for more systematic study of the different factors influencing opinion formation around such technology issues.

(number 4), and each also imported outside wastes (number 5). Large size and the importation of outside wastes, however, were typically regarded as negatively related to the likelihood of siting, so our data do not support those assumptions. Yet only one of the defeated projects, Philadelphia, had definitively excluded the possibility of importing waste. This was a politically sensitive topic, which proponents of such projects preferred to avoid; the question marks indicate defeated sites where it was never clearly admitted by project supporters that trash would be imported, although challengers insisted it was inevitable because of the discrepancies between the sizes of the proposed incinerators and their county waste streams.

Despite the suggestion by industry consultants that population density around a project will be inversely related to the likelihood of siting, our data reveal that the three successful sitings were in some of the more populated areas (number 6) and that in more rural areas, such as Cape May and Broome Counties, projects were defeated. And although the data do not undermine the notion that compensation is necessary to persuade a host community to accept such a plant, specific packages varied considerably across sites, and the total amounts of the compensation showed no obvious relationship to final outcomes (number 7).

The three citizen advisory committees in Delaware, Broome, and St. Lawrence Counties only vaguely resembled their idealized versions portrayed by industry consultants. Of the three projects with CACs, one (Delaware) was a successful siting and two (Broome and St. Lawrence) were defeated (number 8).[3] The notion behind the establishment of a citizen advisory committee is that a small, representative committee of citizens with public credibility and different types of relevant expertise are selected from the broader community to serve as advisers. Ideally, a CAC is involved in the initial site selection processes, in deciding upon plant size, and in all of the other details associated with building an incinerator. We did not find this type of CAC at any of our eight sites, but we did come across a variety of alternatives. In Delaware County, for example, six people were selected by project sponsors to neutralize the Leake administration's opposition to the county plant—and after their customary excursion at the vendor's expense, the group made some minor suggestions for revision in traffic patterns and then ap-

3. Some people in York County, one of the three sites where an incinerator project was successful, referred to a 120-member countywide Solid Waste Advisory Committee as a Citizen Advisory Committee, but this stretches the normal usage of this concept beyond recognition. This York SWAC included representatives from every municipality in the county as well as from environmental groups and manufacturers.

proved a Westinghouse proposal for a site that this CAC had no role in deciding upon. Both of the New York projects also had what some referred to as CACs. In Broome County a group was established "for input on health risk assessments" (according to project proponent Lori Dietz), but opponent Audrey Glover viewed it as "a front for negotiations and a cooptative mechanism." St. Lawrence County's Environmental Management Council (EMC) was the unit some considered a CAC equivalent there, but the last-minute recommendation against the incinerator by this group actually helped undermine that project.[4] One of this EMC's members wrote on our questionnaire:

> The reader must understand that the St. Lawrence EMC was an advisory unit for the St. Lawrence Legislature, and that the incinerator issue was only a small part of our environmental spectrum. It only generated a lot of publicity during the last year of about a five-year span.

These real CACs were quite different from the ideal portrayed by industry consultants. Because of this disparity between what some industry literature recommends and the groups given that label in these and comparable cases, the importance of a genuine CAC in such siting attempts remains to be tested.

Philadelphia was the only one of our eight sites where prior protest mobilization had occurred (number 9). Perhaps the others were intentionally selected to exclude such areas, as the industry report recommended; such an experience is certainly likely to empower local residents. It was easy to recognize a collective feeling of political efficacy in the communities that defeated these projects, but our data do not resolve the empirical question of whether previous protests that have failed or achieved only qualified success promote or inhibit subsequent citizen mobilization.[5]

4. None of the three decisive defeats (Cape May, Lackawanna, Philadelphia) had CACs associated with them, and when asked whether a CAC might have made a difference in Philadelphia, Bruce Gledhill responded: "No . . . It would probably have fed more into City Council's side than into the mayor's. . . . There is a real reluctance on the part of City Council members for the administration to be reaching out into communities because that's their political base. They don't want the administration to be talking to their folks."

5. Neither the projects' actual/estimated costs nor the dates of final decisions revealed noteworthy differences between sitings and defeats. For example, the most expensive project was built, and the least expensive one was defeated. Actual costs, in millions of

Fieldwork Data

The notion that successful sitings are facilitated by having a public champion leading the charge receives partial support from our data, because two of the three successful sitings had such sponsors: Commissioner Catania was critical in Delaware County, and Commissioner Bartle played a similar role in Montgomery County. Yet two of the defeated projects also had their "public champion." Broome County's Carl Young, however, was voted out of office and replaced by Tim Grippen, who opposed the project (hence the Y/N indicator in Table 8 at number 10), and Philadelphia's Mayor Goode was neutralized by his own City Council. The political structure in which the project "champion" operates is important. Where county commissioners use the party machine to get their own way, as was the situation in Delaware and Montgomery Counties, such a "champion" is obviously more critical than in more pluralistic political settings, such as Philadelphia and Broome.

What factors associated with the challenging groups themselves in these disputes made the most difference in affecting outcomes? Relevant data on this issue are found in items 11 to 15 of Table 8. They do not suggest that either the mere emergence of a grassroots social movement organization within a year of a project's first public mention (number 11), or the total number of years spent challenging the project (number 12), was critical in determining outcomes. York's ROBBI, for example, emerged within the first year but was unable to prevent that siting, and Cape May was the most dramatic example of a successful last-minute protest mobilization.

While all of the protest groups considered here had some type of "outside assistance" (number 13)—Delaware County being the single site with no genuine grassroots protest group that outsiders might assist—the nature of such aid varied significantly from project to project. York's ROBBI, for example, received its most important outside support from the new lawyer

dollars, for the three sitings were: $320 (Delaware), $130 (York), and $186 (Montgomery); for the five defeated projects, the estimated costs, again in millions of dollars, were: $79 (Cape May), $120 (Lackawanna), $280 (Philadelphia), $80 (Broome), and $49 (St. Lawrence). And while later final decisions seemed to favor project challengers (see Figure 2), we should keep in mind that the two last-minute defeats in Broome and St. Lawrence Counties might easily have gone the other way. The years for the three sitings were 1987 (York), 1988 (Delaware), and 1989 (Montgomery); those for the five defeats were 1988 (Lackawanna and Philadelphia), 1990 (Cape May and St. Lawrence), and 1992 (Broome).

hired for its litigation, while Broome's backyarders were assisted in their political mobilization by NYPIRG state activists.

Two of the most decisive protest group strategies resulting in defeated projects turned out to be broadening their support base to include grassroots support from outside the backyard area (number 14) and focusing primarily on political rather than legal issues (number 15). Each of the successful protest groups expanded its support base, beyond the particular community where the project was to be constructed, to include groups and individuals from across the county while also focusing primarily on political issues. Some, such as those in Broome and St. Lawrence Counties, used litigation as a supplementary tactic. On the other hand, neither of the two protest groups that formed in backyards where incinerators were built included more than a couple of individuals from outside their backyard communities, and they concentrated primarily on litigation rather than on broader political issues. York's ROBBI and Montgomery County's TRASH, both mobilized to challenge projects eventually sited, had primarily backyard membership bases and focused especially upon litigation in their futile efforts to prevent sitings. Among decisively defeated projects, on the other hand, Cape May's Environmental Response Network, Lackawanna's CARE, and Philadelphia's coalition of neighborhood groups working with City Council members drew upon much wider support bases and focused primarily on political power in their more effective challenges. When asked about "key decisions" and "turning points" for their successful protests, activists at sites where projects were defeated emphasized political activities and outside grassroots assistance with such phrases as "referendum victory," "campaign for Freeholder," "getting the Medical Society's support," "becoming a county-wide coalition."

While political and business leaders typically minimize the influence of grassroots opposition groups on official decisions, spokespersons for such protest organizations also often exaggerate the role of their own social movement organizations in such conflicts. Occasionally, a citizen protest group's demands are taken seriously by some elites but ignored by others, and if this contributes to a split on the issue among these decision-makers it may undermine a project. Keeping in mind the fact that the cleavage between Delaware County and City of Chester authorities was an atypical elite split that did not involve any real anti-incineration faction, the data in Table 8 at number 16 reveal this variable's importance. The other two successful sitings had no splits among political elites, while such splits were indeed present in each of the defeated siting attempts: Cape May's between the new state governor and the Municipal Utilities Authority, Philadelphia's between the mayor and City Council, Broome's between its new county executive and the legislature, St. Lawrence's among county legislators themselves, and Lackawanna's where important political officials supported a protest against a private business venture backed by other county politicians.

The ironic role of the Cerrell Report as a mobilizing tool in these cases casts interesting light on the distinction some make between the so-called "natural" and "social" sciences, and especially on the role of free will in confounding predictions within social science. While our data provided little support for that document's central conclusions, this was partially because the report came into the possession of those whose behavior it presumed to predict and thus contributed to modifying such predictions. This 1984 report suggested that "least resistant" communities would be rural or relatively small, with lower incomes and education, inhabited by older—especially, Catholic—residents, who stood to gain economic benefits from the project. Whatever their validity in the California siting conflicts of the early 1980s from which they were generated, none of these characteristics was useful in distinguishing sitings from defeats among our eight cases in the Northeast. For example, four of the five defeated projects had fewer people living within a two-mile radius than any of the three sited projects, and there were no statistically significant differences in terms of backyarders' incomes, education, or age between sitings and defeats. Counties with larger Catholic populations in the backyard area were more likely to defeat projects, and we have seen that their access to the Cerrell Report itself helped account for this outcome. Activists in fact used this industry forecast of public behavior in mobilizing residents to thwart such predictions.

What can we conclude, therefore, about the factors that promote successful grassroots mobilization? While the complex interactions of details that have an impact upon outcomes in our eight case studies cautions against glib generalizations, some patterns are clearly discernible. Once a project is supported by political and/or economic elites in a county, as was each considered here, widespread grassroots organizing was necessary to stop it. If project challengers were subsequently able to split the elites or to exploit preexisting elite cleavages, they were more likely to frustrate the siting attempt. While litigation was sometimes a useful secondary tactic, this was true only when it was used in conjunction with countywide political organizing.

Before turning to broader theoretical implications of our findings, let us take a final look at Table 8 to compare two specific projects with a view to illustrating critical variables. The siting in Montgomery and the defeat in Lackawanna provide an instructive contrast involving two seemingly similar communities with small percentages of minority residents and virtually identical distributions of residents favoring and opposing their respective projects (see Table 8 at numbers 1, 2, and 3). Looking down the appropriate columns, we find four variables where Montgomery and Lackawanna are identical: both planned to import waste (number 5), neither had even a nominal CAC (number 8), neither had any previous protest (number 9), and both accepted outside assistance (number 13); the two were also similar in size (number 4). Eliminating population density (number 6) and compensation (number 7) from further consideration—because, contrary to predictions, the more densely populated and lower compensated Montgomery County was the place where the incinerator was built—leaves us with six differences between these two communities: Montgomery County had a "political champion" leading the charge for siting (number 10), took longer to mobilize its protest organization (number 11), had a more protracted struggle (number 12), lacked a broad base of support (number 14), focused upon legal rather than political issues (number 15), and did not experience any split among relevant elites involved with the project (number 16). In Lackawanna County, on the other hand, there was no single political leader out front on the incinerator issue, opponents mobilized quickly and established a broad base of grassroots support, they focused primarily upon political rather than legal issues, and they received important assistance from political elites, thus helping neutralize the economic power of project sponsors. Whether this Lackawanna County outcome might have been different had it not been the single site among our eight where an individual entrepreneur rather than the county was the sponsor we cannot say, but these other variables were certainly critical.

Theoretical Reflections

What broader theoretical implications for environmental sociology, and especially for social movements, can we draw from these results?

Environmental Sociology

Whatever the validity of claims about an earlier phase of the environmental movement being elitist, things have changed as grassroots protests have emerged to challenge the impact of modern technologies on public land, air, and water in neighborhoods and communities across the United States and indeed around the globe. When mainstream environmental organizations like the Sierra Club promote direct citizen insurgency—as in Lackawanna County, for example—the professional and grassroots segments of the environmental movement begin to manifest a new synergy, and the movement itself is transformed by its grassroots constituency into what promises to become the dominant social movement in our transition to a new milennium.

Environmental sociology—having persuasively articulated the general link between human social and economic activities, on the one hand, and environmental realities, on the other—is now focusing on more mid-range theory for its continuing development (Freudenburg and Gramling, 1994). Concepts, variables, and hypotheses from contemporary social movement theory are being integrated into emergent new perspectives that span both sociological subfields. Similarities between environmental struggles of minorities against toxic waste sites and the 1960s Civil Rights Movement, for example, promoted the growth of the environmental justice literature (Bullard, 1990). Without denying the abundant evidence of racial injustices in many sitings throughout the United States, it is important to emphasize the complex and changing nature of the processes involved. As systematic comparisons of siting challenges within different types of communities accumulate, we will be able to better understand their dynamics. Not all attempted sitings in minority communities have succeeded, nor, as we have seen here, have they all been unsuccessful when targeted for predominantly white working-class communities. And one of the most fascinating things about research and publication on these topics is that it inevitably has an impact on the phenomena under study.

Social Movement Theory

Contemporary social movement theory will also profit from this rapprochement. There are in that literature a number of central concepts and alleged relationships among variables that should be qualified or questioned on the basis of our data from these examples of grassroots environmental challenges.

One obvious implication for social movement theory from these data is the need for multiple models of mobilization. Distinguishing between traditional equity and more recent technology movements, for example, suggests that relationships among variables that hold for equity struggles such as the civil rights and women's movements may be different for certain types of technological protests, especially those that are in response to suddenly imposed discontents. And even among the latter category of conflicts, additional distinctions are appropriate between grievances deriving from an existing facility, such as an accident at a chemical plant or Three Mile Island, and those precipitated by a proposed plant, such as an incinerator or a waste dump. Challengers intending to eliminate or "displace" their adversaries, for example, typically have better chances for success where they are not confronted by an existing facility. While the evidence from the equity movement literature supports the notion that the surest predictor of failure for a protest organization is the attempt to displace its antagonist, the odds were much better for the technology movements considered here, where five of the eight attempts to displace antagonists were ultimately successful.

The breadth of protest mobilization was a critical variable in accounting for outcomes among these challengers. For this genre of struggle, it is theoretically instructive to compare the effects of organized opposition at different levels for our eight cases (see Figure 3). But a "high" level of organized protest at the local level is not the same as having a relatively high percentage of individual telephone respondents expressing opposition to the incinerator project. Rather, it signifies the presence of coordinated opposition in the backyard community itself. Delaware County's Chester, for example, had no significant grassroots opposition even though 17 percent of its telephone respondents said they opposed the project. And while York's ROBBI was itself a backyard protest organization, it did not really have significant local backing or even the early support of any of its own township officials. Thus, both York and Delaware were classified as "low" for local opposition. While Figure 3 includes three instances of "high" protest mobilization at the local level, one of these—Montgomery County—eventually had the county's incinerator project built despite such opposition. The five cases of "high" countywide organized protest, on the other hand, were all in areas where projects were defeated, regardless of the degree of local protest organization. Even omitting the closely contested New York counties would still leave a clear split between the sited and defeated projects based upon countywide rather than local organized opposition (see Figure 3). These results are relevant in the ongoing debate over the relative importance of indigenous versus outside resources because they suggest that, at least in the case

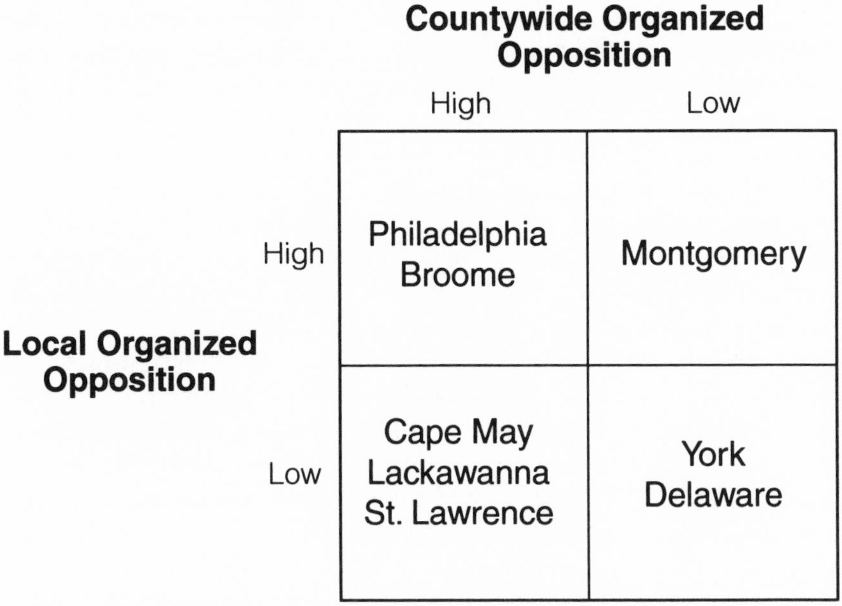

Local Organized Opposition	Countywide Organized Opposition: High	Countywide Organized Opposition: Low
High	Philadelphia Broome	Montgomery
Low	Cape May Lackawanna St. Lawrence	York Delaware

Figure 3. Organized opposition at the local and county levels.

of technology movements involving county projects, indigenous mobilization is less critical than mobilizing surrounding municipalities.

Effective mobilization in these siting struggles was accomplished by concentrating primarily upon political rather than legal issues. Litigation has often been used successfully by equity movements, but that has not been true for technology movements. Where activists relied primarily upon litigation, as in York and Montgomery Counties, they were less successful vis-à-vis both authorities and the grassroots than when they concentrated upon such political activities as petitions, coalition building, referendums, rallies, elections, and lobbying. Besides the fact that even sympathizers are commonly unenthusiastic about contributing money for attorneys in such disputes, the courts and state regulators seldom, if ever, side with citizen challengers. Potential supporters seem more willing to become politically active themselves in challenging such projects when creative protest organizations are available to channel their energies.

While successful political mobilization at the county level required serious framing efforts by the activists, effective framing typically depended upon strategic networking. Opponents not only used anti-incineration master frames—insisting upon prior serious efforts at waste reduction and especially recycling, for example—in structuring their protests at each of the five defeated sites, but also included local amplifications addressed to specific audiences. In Lackawanna County, the anti-incineration master frame was closely linked to the area's negative image as a wasteland—all in the exploitative context of a cynical Cerrell Report. In St. Lawrence County, the anti-incineration master frame was supplemented by a special focus on the shaky economics of the project, intended to persuade the newly elected legislature, and some observers suggested that if WOW had concentrated more on networking with legislators about economic consequences of incineration than on environmental framing they might have won more handily. In Philadelphia, opponents coped successfully with a potentially divisive racial problem that threatened to frame the project as black mayor versus white neighborhoods. In Broome County, the necessary importation of outside garbage to the area persuaded a critical legislator to oppose the project. The "Don't Burn Broome" coalition's work alerting residents to the ramifications of importation apparently had a significant influence on this legislator's decision.

In Cape May County, however, we encountered our most dramatic example of the intimate link between networking and framing. The Environmental Response Network's initial framing efforts had failed to attract the necessary outside support for challenging the incinerator project, and when the protest entrepreneur expanded his own areas of concern to include dolphin deaths, ocean dumping, overdevelopment, and neighborhood problems, regardless of their fit with the group's anti-incineration frame, he alienated co-activists and threatened the ERN's continued existence. On the other hand, these new issues derived from intentionally cultivated networks with other grassroots activists, who subsequently participated in the anti-incineration referendums and election campaigns that eventually derailed the county's project. Instead of manipulating, consciously or unconsciously, the conceptual frame itself so that it was more obviously relevant to the additional issues listed above, as some framing analysts assume must be done, the ERN's Owen Murphy first expanded his networks and volunteered himself with a view toward gaining increased support in the anti-incineration struggle. This suggests that adjusting frames may sometimes be less important than simply expanding activist networks without worrying about framing consistency. These Cape May data show that grassroots activists are sometimes willing to join others in challenging a technology primarily in order to return a favor. There is often a tangled web between framing and networking processes in effective grassroots protests against technological targets.

Our data also suggest that while displacement goals are not as predictive of failure in technology movements as in their equity counterparts, the level of protest mobilization is especially important. The latent strategy for such movements is to neutralize the NIMBY label by expanding their base of support beyond the backyard area. Because the courts regularly defer to political and industry officials rather than to citizen challengers in such disputes, political mobilization rather than a primary focus upon litigation is most promising.

Practical Conclusions in Two Memos

We summarize practical consequences of our findings in the following two hypothetical memos—the first to county officials considering construction of an incinerator, and the second to grassroots citizens opposing it. Some grassroots activists are especially wary of research like ours because, they insist, it is more likely to be used by industry managers than by neighborhood opposition groups. While it may be true that industry has more resources to hire consultants or to pay for systematic literature searches to provide relevant information, there is also increasing evidence on the Internet and elsewhere, that grassroots environmental networks are crisscrossing the United States as well as the world to help local residents challenge LULUs. Industry consultants use a new acronym to refer to this increasingly effective citizen opposition: "BANANA" (Build Absolutely Nothing Anywhere Near Anything). While our specific focus here remains on waste-to-energy incinerators, the memos should also be useful to both sides in similar types of siting conflicts, such as those involving toxic waste facilities, nuclear power plants, nuclear waste dumps, prisons, and other comparable projects.

TO: Relevant county officials
FROM: The authors
RE: Siting projects

Because the pro-industry Cerrell Report has now become an organizing tool for opponents, you should consider this the PC age, where those initials refer not to political language or computers, but rather to a "Post-Cerrell" siting era. While a few plants will probably continue to be built in what that report labels "least resistant" areas, increasing numbers of projects are being defeated by aroused citizen groups identifying themselves sarcastically as a "Cerrell community" and determined to show how much political resistance they really can muster. It is economically irresponsible and politically unwise to risk millions of dollars of citizens' tax money on the questionable siting criteria of the Cerrell Report.

Instead, we suggest that for each project you form a reasonably representative, scientifically competent, and politically independent citizen advisory committee for your initial search. Individuals for the CAC should be selected on the basis of their scientific competence, not for their political connections or previous support for the venture. This CAC should operate from the very beginning—starting with possible site selection and including all other major decisions along the way—and you should make its recommendations available to the general public. Because politics cannot be eliminated from such a process, however, it may be best to notify each municipality in your county about the presumed compensation package associated with such a siting, and then to have the CAC members evaluate potential sites only in areas where elected representatives publicly express interest in hosting the facility.

Even under such conditions, however, you should expect opposition, possibly from a minority of the presumably diverse CAC itself as well as from organized opponents. Try to anticipate the main arguments against the facility by encouraging open discussion ahead of time, and have honest as well as scientifically defensible responses to such objections.

Whether you establish a special county agency for such a project or leave its administration to elected municipal officials, we recommend that you do everything possible to eliminate opportunities for decision-makers to cut deals with vendors and/or subcontractors that neglect the common good. Be wary of vendor-sponsored trips to exotic locations, which are commonly interpreted as bribes. Even the suspicion that county officials have made self-aggrandizing decisions in spending public funds promotes grassroots opposition.

Assuming that you would not initiate such a project until you are sure that you have the support of county legislators or city council members voting on it or its bonds, you should also be willing to accept additional delays, which promise to promote a more informed decision. Realize the weakness of arguments based upon how much has already been mistakenly spent on a project, and recognize that time is not necessarily your enemy. Whether it is the discovery of cheaper and more efficient alternative technologies (for example, recycling to reduce the trash stream) or the discovery of flaws in proposed solutions (for example, problems with dioxins and residual ash), delayed decisions may prevent a major health hazard and/or financial blunder for your county.

Finally, do not mistake acceptance of a project by the prospective backyard community as a guarantee of a successful siting. Remember Ogdensburg!

. . .

TO: Grassroots opponents
FROM: The authors
RE: Challenging county sitings

Frame your initial objection in a countywide rather than local context. Organizations such as the Citizens Clearinghouse for Hazardous Wastes (CCHW) are valuable sources of substantive information about recycling and toxic wastes and about strategies for using such information. Draw heavily upon existing networks, not only to learn more about critical issues but also as sources of emotional support from those with similar experiences.

Focus primarily upon political rather than legal concerns. Enlist mainstream group representatives early in opposing the project. Grassroots mobilization is more likely than litigation alone to build rapport among potential supporters. Litigation typically alienates people and reduces participation as money is collected to pay distant attorneys with little real passion for your cause. While litigation may sometimes be useful in delaying these projects or harassing their sponsors, it is seldom effective except in conjunction with political organizing—which may also pressure the courts, where judges typically rule in favor of state and county authorities in disputes with citizens.

If the proposed project has a political champion, try to personalize the dispute by focusing upon that individual. For example, show what that person (or small group) stands to gain financially from the project—in contrast to its public costs in terms of recycling counterincentives, threats to health and safety from pollutants and ash, lowering property values, increased truck traffic, noise, and so on. Use every means possible to get your message to the audience of the uncommitted—for example, demonstrations in strategic locations where television coverage is assured, letters to the editor, public debates, and referendums (even nonbinding referendums as in Cape May).

Do not allow your group to be portrayed by the press or proponents as NIMBYs or, worse, radicals, because that threatens your potential base of political support among mainstream audiences. If some of your sympathizers insist on more militant action in pursuit of the common goal, publicly dissociate your protest organization from them, utilizing "the function of the radical fringe" by portraying your group as moderate in comparison.

While working to continually expand the grassroots base of your group's political support, also seek to identify and recruit important decision-makers who are sympathetic to your position. Consider promoting a publicly independent splinter group to lobby legislators and other potential mainstream supporters. Be alert to splits, even potential ones, among the relevant decision-makers, realizing that an ideal political situation for your purposes is to have elites publicly disagreeing over the focal issues.

Give your own members plenty of political space because such grassroots struggles tend to take over and disrupt lives, commonly leading to "burn out" and personal confusion. Allow people to go AWOL from time to time, always welcoming them back with no questions asked. Instead of adopting the common movement outlook that "s/he who is not for us is against us," cultivate the more creative "s/he who is not against us is (at least potentially) for us."

Finally, don't lose heart. Become increasingly creative in your strategies and tactics, even when it seems you are standing alone. Remember Owen Murphy!

Appendix: Sampling and Statistical Methods

Telephone interviews were conducted with 800 randomly selected respondents ("backyarders") from the eight sites. A professional telephone interviewing firm identified backyarders by calling households in telephone exchanges near or in the area where the incinerator was to be located. Only people 18 years or older were interviewed. An approximately equal number of men and women were interviewed (see Table A.1). Two filter questions were asked to locate the backyarders precisely. The first question was "Have you lived in the (name of site) area for (two) three years or more?" The time varied depending on the site, but covered the time the incinerator conflict was in progress. If respondents answered yes, they were asked either "About how far is the incinerator in (name of site) from your home?" or "About how far would the proposed incinerator in (name of site) have been from your home?" Those who answered three miles (30 blocks) or fewer were considered to be backyarders.

We were unable to use a full random-digit dialing procedure because of limited resources. Telephone numbers were randomly selected from the telephone book, and the last two numbers were replaced by two numbers drawn at random. The cooperation rates for the eight sites—that is, those who agreed to be interviewed of the total number contacted, are as follows:

Broome County (N.Y.)	76%
Cape May County (N.J.)	67%
Delaware County (Pa.)	60%
Lackawanna County (Pa.)	65%
Montgomery County (Pa.)	62%
Philadelphia (Pa.)	55%

St. Lawrence County (N.Y.)	74%
York County (Pa.)	73%

In all but two sites—Broome and St. Lawrence—the backyarders were interviewed after the incinerator controversy had been definitely decided. The dates of the project outcomes and the dates of the backyarder interviews are as follows:

Delaware	Construction began December 1988 Interviews conducted July 1990
Montgomery	Construction began May 1989 Interviews conducted November 1990
York	Construction began October 1987 Interviews conducted November 1990
Broome	Project abandoned February 1992 Interviews conducted October 1990
Cape May	Project abandoned July 1990 Interviews conducted October 1990
Lackawanna	Project abandoned December 1988 Interviews conducted October 1990
Philadelphia	Project abandoned July 1988 Interviews conducted June 1990
St. Lawrence	Project definitively abandoned in 1991 Interviews conducted December 1990

Three statistical tests were used for the data displayed in Tables 1, 2, and 3 and Table A.1 in this Appendix. Chi-square tests were used to compare the difference between the distributions reported as "percentages." The differences between two means and three or five means were evaluated by t-tests and F-tests respectively.

The Survey Questions

The exact questions the backyarders were asked are presented in this section, divided into two groups. The first set of questions are related to Tables

1, 2, and 3. The variable numbers on those tables correspond to the question numbers below. For example, variable number 1 is "Percent opposed to siting of trash plant," and question 1, below, provides the reader with a full description of the question upon which variable 1 is based—that is, "How did you feel at the time about the (name of site) trash plant? Were you in favor, opposed, did you have mixed feelings, or didn't you have any opinion on the issue?" The response category "Opposed" is highlighted in **bold** to indicate that the percentage is based on the number opposed divided by the total number of all responses.

The "Attitudes Toward Incinerator Plants" items are framing and ideology measures, based on statements by proponents and opponents from interviews, newspaper accounts, and other publications that came to the attention of the authors. The attitude questions were asked in the present tense for the backyarders who were interviewed where plants were sited, and in the past tense for backyarders who lived in areas where the plant was defeated. The network and background items were based upon the literature summarized in Chapter 2.

Survey Questions Related to "Attitudes Toward Incinerator Plants" (Framing and Ideology Measures)

1. How did you feel at the time about the (name of site, e.g., Cape May) trash plant? Were you in favor, opposed, did you have mixed feelings, or didn't you have any opinion on the issue?

   Favor, Mixed Feelings, **Opposed,** No Opinion

2. How much will the new plant help solve (name of site) trash problems? or How much would the trash plant have helped solve (name of site) trash problems? Will it help (would it have helped) . . .

   Very much, Pretty much, Not very much, Don't know

3. A trash plant is good for a community's economy. Do you . . .

   Agree, Disagree, Don't know

4. Burning trash in a trash plant is better than dumping it into a landfill. Do you . . .

   Agree, Disagree, Don't know

5. Trash plants are bad for human health. Do you . . .

   Agree, Disagree, Don't know

6. How likely is it that the plant will lower your property values or hurt you financially? or How likely was it that the plant would have lowered your property values or hurt you financially? Is it (was it) . . .

   Very likely, Somewhat likely, Not very likely, Don't know

7. If a community builds a trash plant, then it won't make serious efforts to recycle its trash. Do you . . .

   Agree, Disagree, Don't know

Survey Questions Related to "Network Characteristics"

8. Do you remember discussing the (name of site) trash plant with any family or friends during the time when it was in the news?

   Yes, No, Don't know

9. With how many different people did you discuss it?

   Actual number

10. Did any of your friends or neighbors feel differently than you about the (name of site) trash plant?

    Yes, No, Don't know

11. Did you hear of any public meeting in your area on the trash plant issue?

    Yes, No, Can't recall

12. Did you attend any such public meeting?

    Yes, No, Can't recall

13. During that time, that is (dates of conflict), did you belong to any organizations or other groups that were against the trash plant?

    Yes, No, Can't recall

14. During that time, did you belong to any organizations or groups that were not involved in the incinerator or trash plant issue?

    Yes, No, Don't recall

15. How much information about the trash plant issue did you get from conversations with other people? Would you say you got . . .

    Quite a bit of information, Some information, Hardly any or no information, Don't know

Survey Questions Related to "Background Characteristics"

16. During the 1988 presidential election, did you vote for Bush, Dukakis, someone else, or didn't you vote?

    Bush, Dukakis, Someone else, Didn't vote, Refused

17. What is the highest grade of school you completed?

    Graduate/Professional, **Completed college**, **Some college**, Completed high school, Some high school, Grade school, None, Refused

18. Would you tell me, what was the total income of your entire household before taxes last year?

    Over $35,000, **Between $25,001 and $35,000**, Between $15,001 and $25,000, Between $10,001 and 15,000, $10,000 or less, Don't know, Refused

19. Do you own your residence or are you renting?

    Own (or buying), Renting, Living in someone else's home, Living in an institution or group home, Other, Refused

20. Would you say you are . . .

    Catholic, Protestant, Jewish, Other, No preference or None, Refused

21. Would you describe yourself as . . .

    White, Black, Hispanic, Oriental, Other, Refused

22. How long have you lived in (name of site)?

Number of years

23. What is your present age?

Number of years

The additional questions, starting on page 270, were also included in the survey of backyarders but not presented in the tables in Chapter 3. In most cases, these questions were similar in content to the questions discussed in Chapter 3 or did not differentiate backyarders who lived where plants were sited from those who lived where the plant was not sited. Table A.1 presents a summary of these data, and the numbers in the table correspond to the question numbers.

In the first section of Table A.1 are attitude questions about the plant. Variable numbers 2, 3, and 4 are similar to variable numbers 4, 5, and 7 in Table 1. The differences are similar to those in Table 1. It may be noted that a high percentage of those interviewed were aware of the siting conflict. Some 95 percent of those living in the communities where the plant was defeated, and 89 percent of those living where the incinerator was sited, were aware of the conflict.

The second section of Table A.1 demonstrates that the two groups have similar experience with previous protest movements. The third section indicates that the groups also do not differ on other background characteristics not mentioned in Chapter 3.

The last section reveals two significant differences. Those living where the plant was defeated were more likely to get information about the siting conflict from television news and from radio. These media findings are explored in more detail in Chapters 4 through Chapter 7.

Table A.1 Additional comparisons of backyarder characteristics

Backyarder characteristics	Sited	Defeated
Attitudes toward incinerator plants		
1. Percent who were aware of siting conflict	89.0	95.0*
2. Percent who believed recycling "very important" solution to trash problem	89.0	87.0
3. Percent who believed incinerators threat to community health and safety	54.0	73.0**
4. Percent who agreed incinerators reduce landfill wastes	91.0	80.0**
5. Percent who trust scientific studies on health hazards of incinerators	56.0	51.0
Previous protest involvement		
6. Percent who contributed time or money to Civil Rights Movement	14.0	12.0
7. Percent who contributed time or money to Anti-Vietnam War Movement	7.0	8.0
8. Percent who contributed time or money to Women's Movement	9.0	10.0
9. Percent who contributed time or money to Anti-Nuclear Movement	5.0	6.0
10. Percent who contributed time or money to Prolife Movement	9.0	13.0
11. Percent who were involved in local protests	28.0	32.0
Background characteristics		
12. Percent who are female	49.0	52.0
13. Percent who have five or more friends living within a mile	75.0	77.0
14. Percent who are employed full time	48.0	53.0
15. Percent who are married	66.0	61.0
Media sources		
16. Percent who got "quite a bit of information" about the plant from newspapers	44.0	47.0
17. Percent who got "quite a bit of information" about the plant from TV news	9.0	28.0**
18. Percent who got "quite a bit of information" about the plant from radio	4.0	16.0**

*$p < .05$
**$p < .01$

Survey Questions Related to "Attitudes Toward Incinerator Plants" (Framing and Ideology Measures)

1. Do you remember all the talk about putting a trash plant in (name of site)?

    Yes, No, Can't recall

2. How important is it to recycle trash to help solve the trash problem? Would you say. . .

    Very important, Somewhat important, Not very important, Don't know

3. How much of a threat will (would) the trash plant be (have been) to the health and safety of people in the area? Will it be (Would it have been) . . .

    A very serious threat, Somewhat of a threat, Not a very serious threat, Don't know

4. Trash plants cut down the amount of waste that goes into landfills. Do you . . .

    Agree, Disagree, Don't know

5. You can trust scientific studies about how trash plants affect human health. Do you . . .

    Agree, Disagree, Don't know

Appendix Survey Questions Related to "Previous Protest Involvement"

6. Have you ever contributed any time or money to the Civil Rights Movement?

    Yes, No, Don't recall

7. Have you ever contributed any time or money to the Anti-Vietnam War Movement?

    Yes, No, Don't recall

8. Have you ever contributed any time or money to the Women's Movement?

    Yes, No, Don't recall

9. Have you ever contributed any time or money to the Anti-Nuclear Movement?

    Yes, No, Don't recall

10. Have you ever contributed any time or money to the Anti-Abortion Movement?

    Yes, No, Don't know

11. In the past, have you ever been involved in any local protest in your community?

    Yes, No, Can't recall

Appendix Survey Questions Related to "Background Characteristics"

12. Gender of Respondent.

    Female, Male

13. How many friends live within a mile or so of your home? Would you say . . .

    Five or more, Three or four, One or two, None, No idea

14. Are you employed full time, part time, or not employed?

    Full time, Part time, Not employed, Refused

15. Are you . . .

    Married, Separated, Divorced, Widowed, Never Married, Refused

Appendix Survey Questions Related to "Media Sources"

16. How much information about the trash plant issue did you get from newspapers? Would you say you got . . .

    Quite a bit of information, Some information, Hardly any or no information, Don't know

17. How much information about the trash plant issue did you get from TV news? Would you say you got . . .

    Quite a bit of information, Some information, Hardly any or no information, Don't know

18. How much information about the trash plant issue did you get from radio stations? Would you say you got . . .

    Quite a bit of information, Some information, Hardly any or no information, Don't know

References

Anastasia, George. 1993. "Pa. Crime Commission Ready to Cap Its Poisoned, Pithy Pen." *Philadelphia Inquirer*, December 19, p. C1.

Anderson, Kelvyn. 1990. "Trash Plant Draws Fire." *Delaware County Sunday Times*, April 22.

Apter, Jennie. 1992. "Trash Importation Vote Stalls Incinerator: County Legislature Fails to Override Veto by One Vote." *Pipe Dream* (SUNY Binghamton), February 21.

Argento, Michael. 1987. "150 on Hand for DER's Hearing on Incinerator." *York Daily Record*, January 13.

Banchero, Stephanie. 1991. "At Last, Trash Becomes Steam." *Philadelphia Inquirer*, November 24.

Basler, George. 1987. "100 Call for Trash Burning Plan Ban." *Broome County Press & Sun-Bulletin*, May 3.

Blumberg, Louis, and Robert Gottlieb. 1989. *War on Waste: Can America Win Its Battle with Garbage?* Washington, D.C.: Island Press.

Books, Kenneth. 1988. "Massachusetts Incinerator Impresses City's Engineer." *Scrantonian Tribune*, September 14.

Bradford, Peter. 1979. "The Nuclear Option: Did It Jump or Was It Pushed?" Manuscript of a talk delivered in East Lansing, Michigan, August 2.

Brooks, Tad. 1985. "If There's Smoke, There Can Be Dioxin." *Norristown Times Herald*, August 16.

Bukro, Casey. 1989. "East's Garbage Filling Heartland." *Chicago Tribune*, November 5, pp. 1ff.

Bullard, Robert D. 1990. *Dumping in Dixie: Race, Class, and Environmental Quality.* Boulder, Colo.: Westview Press.

Bullard, Robert, and Beverly Wright. 1992. "The Quest for Environmental Equity: Mobilizing the African-American Community for Social Change." In Riley Dunlap and Angela Mertig, eds., *American Environmentalism.* New York: Taylor & Francis.

Buttel, Fredrick. 1987. "New Directions in Environmental Sociology." *Annual Review of Sociology* 13:465–88.

Cable, Sherry, and Charles Cable. 1995. *Environmental Problems, Grassroots Solutions: The Politics of Grassroots Environmental Conflict.* New York: St. Martin's Press.

Cable, Sherry, Ed Walsh, and Rex Warland. 1988. "Differential Paths to Political Activism: Comparisons of Four Mobilization Processes After the Three Mile Island Accident." *Social Forces* 66:951–69.

Cadden, James. 1988. "Incinerator Sparks Mass Demonstration." *Scrantonian Tribune*, August 17.

Caldwell, Lynton. 1992. "Global Environmentalism: Threshold of a New Phase in International Relations." In Riley Dunlap and Angela Mertig, eds., *American Environmentalism*. New York: Taylor & Francis.

Cartier, Cathy. 1990a. "O'Connor: Conditions May Not Kill Incinerator." *Ogdensburg Journal*, June 11.

———. 1990b. "EMC Agrees to Rethink Position on Incinerator; Set July 5 Session." *Plaindealer*, June 27.

Cartier, Cathy, and Jim Reagen. 1990. "Legislature Poll Shows Incinerator Passing." *Ogdensburg Journal*, June 29.

Cass, Julia. 1986. "Strong Opposition in S. Phila." *Philadelphia Inquirer*, June 26.

CCHW (Citizens Clearinghouse for Hazardous Wastes Inc.). 1986. *Five Years of Progress, 1981–1986*. Arlington, Va.: Citizens Clearinghouse for Hazardous Wastes.

Cerrell Associates Inc. 1984. Political Difficulties Facing Waste-to-Energy Conversion Plant Siting. Los Angeles: California Waste Management Board.

Clark, Robin. 1987. "Panel Tables Goode's Trash Plan." *Philadelphia Inquirer*, January 29.

———. 1988. "It Was Familiar Opposition to a Familiar Proposal." *Philadelphia Inquirer*, April 14.

Coffin, Tristram. 1992. "From the Files." *Washington Spectator*. New York: Public Concern Foundation, August 1, 1992.

Collette, Will. 1989. "The Polluters' 'Secret Plan.'" Arlington, Va.: Citizens Clearinghouse for Hazardous Wastes.

Commoner, Barry. 1990. "Ending the War Against the Earth." *The Nation* 250, April 30.

Compart, Andrew. 1987. "Incinerator Work Starts in Three Weeks." *York Daily Record*, July 27.

Connett, Ellen, and Paul Connett. 1993. "Waste Not," no. 251, November.

Conroy, Jim. 1988. "Famed Environmentalist Here to Help Block Bechtel Plant." *Scranton Times*, September 14.

Conroy, Theresa. 1986. "Trash-to-Steam Foes Testify at Hearing." *Philadelphia Inquirer*, October 23.

Cooke, Russell. 1985. "Chamber Turns Against Goode on Trash Plan." *Philadelphia Inquirer*, February 26.

Cummings, William. 1989a. "Judge Refuses to End Probe: Incinerator Hearings Delayed." *Watertown Daily Times*, June 28.

———. 1989b. "SWDA Shows Sign of Rift: Vote on Landfill Option Casts Doubt on Incinerator." *Watertown Daily Times*, May 17.

———. 1989c. "Cancer Risk Assessment Drastically Lowered." *Watertown Daily Times*, June 6.

———. 1989d. "Connett Calls for State Inquiry into Health Risk Adjustments." *Watertown Daily Times*, June 15.

———. 1989e. "SWDA Digs Up Dirt on Witness." *Watertown Daily Times*, October 6.

Curlee, T. Randall, Susan M. Schexnayder, David P. Vogt, Amy K. Wolfe, Michael P. Kelsay, and David L. Feldman. 1994. *Waste-to-Energy in the United States: A Social and Economic Assessment*. Westport, Conn.: Quorum Books.

Curran, Robert. 1988a. "Incinerator Engulfed in Smog of Words." *Scrantonian Tribune*, February 25.

———. 1988b. "Bechtel Burner Under Fire." *Scrantonian Tribune*, August 27.

———. 1988c. "Sierra Questions Bechtel Motives in Semass Tour." *Scrantonian Tribune*, September 14.

———. 1988d. "Medicos Urge Burner Moratorium." *Scrantonian Tribune*, September 17.

———. 1988e. "HELP: Planned Incinerator 'One More Nail in Region's Coffin.'" *Scrantonian Tribune*, September 16.

Davis, Andy. 1988a. "ROBBI Observes 2nd Anniversary." *York Dispatch*, February 24.

———. 1988b. "Ash Disposal Is Discussed." *York Dispatch*, March 18.

Degener, Richard. 1990. "Cape Won't Burn Trash in Area; May Send It to Pa. or Compost It." *Atlantic City Press*, May 8.

Degener, Richard, and Kathleen Cannon. 1988. "County Voters Overwhelmingly Reject Mass Burn." *Atlantic City Press*, November 9.

Demme, David. 1993. "Gaining Ground in the Future?" *Municipal Solid Waste Management*, November.

Denison, Richard A., and John Ruston. 1990. "Introduction." In Richard A. Denison and John Ruston, eds., *Recycling and Incineration*, pp. 3–26. Washington, D.C.: Island Press.

Donnelly, James R. 1990a. "Delay May Cost Authority $3.5M." *Watertown Daily Times*, June 13.

———. 1990b. "Say No to Burner, Firm Urges in Md.: Cites Advantages of SWDA's Project." *Watertown Daily Times*, June 30.

Dunlap, Riley, and Angela Mertig. 1992. "The Evolution of the U.S. Environmental Movement from 1970 to 1990: An Overview." In Riley Dunlap and Angela Mertig, eds., *American Environmentalism*. New York: Taylor & Francis.

Eib, Lynn. 1986. "Incinerator Foes: Landfill Owner Joins Citizens Group to Oppose County Project." *York Daily Record*, May 28.

Emery, Margaret. 1988a. "Dalton Not Eager to Join Incinerator." *Scranton Times*, August 5.

———. 1988b. "Dunmore, DeNaples Family Agree to Not Build Incinerator at Landfill." *Scranton Times*, September 29.

Environmental Protection Agency (EPA). 1988. "Report to the U.S. Congress: Solid Waste Disposal in the United States," vol. 2, EPA/530-SW-88-011B (Washington, D.C., October).

Erikson, Kai. 1991. "A New Species of Trouble." In Stephen Couch and J. Stephen Kroll Smith, eds., *Communities at Risk: Collective Response to Technological Hazards*. New York: Peter Lang.

Eshleman, Russell E. 1986. "To S. Philadelphians, Trash Plan Stinks." *Philadelphia Inquirer*, April 28.

———. 1991. "Casey Trash Plan Likely to Draw Fire." *Centre Daily Times* (State College, Pa.), January 8, p. 1.

Fazlollah, Mark. 1991. "Grand Jury Accusations Deal Chester Another Blow." *Philadelphia Inquirer*, April 14.

Frassinelli, Mike. 1987a. "DER Gives Incinerator Plans Green Light." *York Daily Record*, May 14.

———. 1987b. "County Incinerator Bonds Go on Sale." *York Daily Record*, August 7.

Freudenburg, William, and Robert Gramling. 1994. "Mid-Range Theory and Cutting-Edge Sociology: A Call for Cumulation." *Environment, Technology, and Society* (newsletter of the American Sociological Association Section on Environment and Technology), no. 76 (Summer): 1–7.

Gailey, Tom. 1986. "Bartle Firm on Barring Phila. Trash." *Norristown Times Herald*, February 22.

Gamson, William A. 1980. "Understanding the Careers of Challenging Groups: A Commentary on Goldstone." *American Journal of Sociology* 85:1043–60.

Gerhards, Jurgen, and Dieter Rucht. 1992. "Mesomobilization: Organizing and Framing in Two Protest Campaigns in West Germany." *American Journal of Sociology* 98 (November): 555–95.

Gibbons, Margaret. 1985a. "County Incinerator Seen by Year End." *Norristown Times Herald*, February 22.

———. 1985b. "County Slates 2nd Hearing on Trash Plant." *Norristown Times Herald*, August 7.

———. 1985c. "County Commits to Construction of Trash Plant." *Norristown Times Herald*, December 12.

———. 1985d. "Opponents Lash Out at Bartle." *Norristown Times Herald*, November 22.

———. 1986a. "DER Review Challenged by Plymouth." *Norristown Times Herald*, April 17.

———. 1986b. "Judge Gives Go-Ahead for Trash Plant." *Norristown Times Herald*, August 11.

———. 1987. "County, Plymouth Face June Court Date." *Norristown Times Herald*, April 3.

———. 1988a. "Trash Plant Gains: Plymouth Ordered to OK Land Plan." *Norristown Times Herald*, June 18.

———. 1988b. "Dravo Lines Up New Firm to Build Plymouth Incinerator." *Norristown Times Herald*, July 26.

———. 1988c. "County Claims Major Win in Trash Plant Battle." *Norristown Times Herald*, October 21.

———. 1988d. "County Sticks with Incinerator Plan." *Norristown Times Herald*, November 24.

Gibbons, Michael. 1986a. "Meeting on Proposed Plant Becomes 6-Hour Marathon." *Norristown Times Herald*, October 22.

———. 1986b. "For Council Chief, Who's Talking About Trash Plant Important as Who Isn't." *Norristown Times Herald*, November 22.

Gleason, Jerry L. 1988. "Foes of Incinerator Hope to Halt Building of York County Unit." *Harrisburg Sunday Patriot-News*, January 10.

Goldstone, Jack. 1980a. "The Weakness of Organization: A New Look at Gamson's *The Strategy of Social Protest.*" *American Journal of Sociology* 85:1017–42.

———. 1980b. "Mobilization and Organization: Reply to Foley and Steedly and to Gamson." *American Journal of Sociology* 85:1428–32.

Grady, Peter. 1988. "Incinerator Suffers 2-Pronged Attack." *Scrantonian Tribune*, August 31.

Grippen, Timothy M. 1988. "Citizens Right to Fight Incinerator Plan." *Broome County Press & Sun-Bulletin*, August 17.

Hart, Joe. 1986. "Opponent Rips County Trash Plant." *Delaware County Daily Times*, November 28.

———. 1989. "Panel Links Ex-Mayor to Trash Plant." *Delaware County Daily Times*, February 27.

Hocker, Christopher. 1991. "Waste-to-Energy Development: Who's Doing What and Why?" *Solid Waste and Power*, August, 12–19.

Jasper, James M., and Jane D. Poulsen. 1995. "Recruiting Strangers and Friends: Moral Shocks and Social Networks in Animal Rights and Anti-Nuclear Protests." *Social Problems* 42(5): 493–512.

Jenkins, Patrick. 1987. "DEP Declares War on Conspirators of 'Not in My Backyard' Syndrome." *Newark Star-Ledger*, June 19.

Kennedy, Sara. 1986. "The Path of Plans for the City's Trash Has Many Twists and Turns." *Philadelphia Inquirer*, April 27.

Klandermans, Bert. 1984. "Mobilization and Participation: Social Psychological Expansions of Resource Mobilization Theory." *American Sociological Review* 49(5): 583–600.

———. 1986. "New Social Movements and Resource Mobilization: The European and American Approach." *International Journal of Mass Emergencies and Disasters* 4(2): 13–39.

———. 1988. "The Formation and Mobilization of Consensus." In Bert Klandermans, Hanspeter Kriesi, and Sidney Tarrow, eds., *From Structure to Action: Comparing Movement Participation Across Cultures*, 173–97. *International Social Movement Research*, vol. 1. Greenwich, Conn.: JAI Press.

Knoke, David. 1990. *Political Networks: The Structural Perspective*. New York: Cambridge University Press.

Konheim, Carolyn S. 1986. "Reconciling Local Opposition to a Resource Recovery Plant." *New Jersey Bell Journal* 9:45–55.

LaBarth, Len. 1988a. "Westinghouse Is Sure Its Proposal Faces Fire." *Delaware County Daily Times*, April 26.

———. 1988b. "Hundreds Join Forces to Battle Trash Plant." *Delaware County Daily Times*, April 28.

———. 1988c. "Delco's Project Gathers Steam." *Delaware County Daily Times*, August 1.

———. 1988d. "Westy Gets Go-Ahead." *Delaware County Daily Times*, September 25.

LaBarth, Len, and Adam Taylor. 1988. "Delco Threatens to Stop Taking City Trash." *Delaware County Daily Times*, May 5.

Lalli, Steve. 1988a. "Councilmen Won't Be Bulled on Incinerator Decision." *Scranton Times*, August 2.

———. 1988b. "Mass. Visitors May Meet Incinerator Foes." *Scranton Times*, September 10.

Lapusheski, Christine. 1990a. "Solutions on Trash Offered—Three Companies Vying to Compost, or Burn." *Atlantic City Press*, May 24.

———. 1990b. "Cape Picks Firm for Waste Disposal." *Atlantic City Press*, July 19.

Lau, Edie. 1990. "State Chief Hears Burner Arguments: 'Messy' Issue Needs 'Extraordinary Steps,' Jorling Says." *Press & Sun-Bulletin*, August 16.

———. 1991. "Burner Changes Ordered: Broome Must Import Trash or Cut Plant Size." *Press & Sun-Bulletin*, December 19.

Lawson, Gregg. 1987. "Environmentalists Form 'Network.'" *Cape May Herald-Lantern-Dispatch*, April 15.

League of Women Voters. 1993. *The Garbage Primer*. New York: Lyons & Buford.

Loeb, Vernon. 1985. "Starting Over: A Long Road for Goode's New Trash Plan." *Philadelphia Inquirer*, January 27.

Lovett, Kenneth. 1989. "Vote Heartens Incinerator Foes." *Watertown Daily Times*, May 17.

Lukowski, Stan. 1988. "Dickson Plans DER Protest on Incinerator." *Scrantonian Tribune*, March 3.

Lyon, Paul. 1988a. "Incinerator Builder Differs with Environmentalists." *Scranton Times*, September 16.

———. 1988b. "Huge Crowd Attends Incinerator Hearing." *Scranton Times*, September 21.

Manly, Howard. 1985a. "Opposition Vocal, but Council Votes for Trash District." *Philadelphia Inquirer*, April 18.

———. 1985b. "11 Towns Agree in Principle to Allow County Trash Control." *Philadelphia Inquirer*, April 15.

———. 1985c. "To Promote Plymouth Trash Plant, Developer Invites 60 to Baltimore." *Philadelphia Inquirer*, May 13.

———. 1985d. "Council Reverses Again, Offers 4th Trash Plan." *Philadelphia Inquirer*, July 1.

———. 1985e. "Montco Trash-to-Steam Plant Has Cleared Some Hurdles." *Philadelphia Inquirer*, December 30.

Martin, Ryne R. 1989. "New Legislature to Decide on Incinerator." *Ogdensburg Journal*, December 12.

———. 1990. "Legislators Getting Cold Feet as Time Draws Near for OK." *Ogdensburg Journal*, June 12.

Marx, Gary T., and James L. Wood. 1975. "Strands of Theory and Research in Collective Behavior." *Annual Review of Sociology* 1:363–428.

McAdam, Doug. 1982. *Political Process and the Development of Black Insurgency, 1930–1970*. Chicago: University of Chicago Press.

———. 1986. "Recruitment to High-Risk Activism: The Case of Freedom Summer." *American Journal of Sociology* 92:64–90.

McAdam, Doug, John McCarthy, and Mayer Zald. 1988. "Social Movements." In Neil Smelser, ed., *Handbook of Sociology*, pp. 695–737. Beverly Hills, Calif.: Sage.

McCaffrey, Eileen. 1985a. "Looking to TRASH Zoning Plan: Residents Protest Bid to Locate Incinerator in Plymouth." *Norristown Times Herald*, April 23.

———. 1985b. "Plymouth Citizens' Group Persists in Bid to Stop Trash-to-Fuel Plant." *Norristown Times Herald*, April 26.

———. 1985c. "Protest Not Linked to Politics, Says Leader of Trash Plan Foes." *Norristown Times Herald*, April 27.

———. 1985d. "Trash District Is Repealed by Plymouth." *Norristown Times Herald*, April 30.

———. 1985e. "Conshohocken Council Staying Away from Trash-Plant Issue." *Norristown Times Herald*, June 20.

———. 1986. "Citizens' Group Chastises Plymouth Council on 2 Trash-Burning Issues." *Norristown Times Herald*, April 15.

McCarthy, John, and Mayer Zald. 1977. "Resource Mobilization and Social Movements: A Partial Theory." *American Journal of Sociology* 82:1212–41.

McCarthy, T. J. 1987a. "MUA Says Trash-to-Ash Dioxin Risk Is Negligible." *Atlantic City Press*, February 12.

———. 1987b. "Opponents Call Mass Burn 'A Dangerous Solution.' " *Atlantic City Press*, March 7.

———. 1987c. "Mass-Burn Ash: Landfill-Safe or Toxic Waste?" *Atlantic City Press*, March 9.

———. 1987c. "Mass Burn Called Integral to State Solid Waste Plan." *Atlantic City Press*, August 27.

———. 1987d. "County Approves Mass-Burn Plant." *Atlantic City Press*, September 18.

McCloskey, Michael. 1992. "Twenty Years of Change in the Environmental Movement: An Insider's View." In Riley Dunlap and Angela Mertig, eds., *American Environmentalism*, pp. 77–88. New York: Taylor & Francis.

McCormack, Kathy. 1989. "Incinerator Challenge Heads to Court." *York Dispatch*, September 5.

McCullough, Michael. 1987. "Demonstrators, Speakers Protest Sludge Dumping, Pollution." *Atlantic City Press*, September 7.

McKenna, Patrick. 1988. "Ash Safe, Local Visitors to Incinerator Told." *Scranton Times*, September 15.

McLaughlin, Harry. 1986. "Opponent of York Incinerator Warns Residents of Cancer Risk." *Harrisburg Patriot*, October 21.

Melosi, Martin. 1981. *Garbage in the Cities: Refuse, Reform, and the Environment, 1880–1980*. College Station: Texas A&M University Press.

Miller, Bill. 1988. "Trash-to-Steam Is Denounced at City Hall Rally." *Philadelphia Inquirer*, April 17.

Molotch, Harvey. 1979. "Media and Movements." In Mayer N. Zald and John D. McCarthy, eds., *The Dynamics of Social Movements*, pp. 71–91. Cambridge, Mass.: Winthrop Publishers.

Montaigne, Fen. 1986. "S. Phila. Crowd Assails Proposal for Trash Plant." *Philadelphia Inquirer*, November 11.

Morris, Aldon D. 1992. "Political Consciousness and Collective Action." In Aldon D. Morris and Carol McClurg Mueller, eds., *Frontiers in Social Movement Theory*, pp. 351–74. New Haven, Conn.: Yale University Press.

Morrison, Denton. 1986. "How and Why Environmental Consciousness Has Trickled Down." In Allan Schnaiberg and Nicholas Watts, eds., *Distributional Con-*

flicts in Environmental Resource Policy, pp. 187–220. Aldershot, Eng.: Gower Publishing.

———. 1989. "The Environmental Movement in the United States: A Developmental and Conceptual Examination." Working Paper, Michigan State University.

Musselwhite, Cheryl. 1991. "Demonstrators Rally Against County Incinerator." *Pipe Dream* (SUNY Binghamton), October 25.

Odato, James M. 1988. "Anti-Incinerator Groups Say They'll Sue Broome on Solid Waste Plan." *Broome County Press & Sun-Bulletin*, July 24.

Palfrey, David. 1988. "Jefferson to Discuss Dunmore Incinerator Plan." *Scranton Times*, August 1.

Pfitzer, Kurt. 1985. "Proposal Would Allow Incinerator for Area Use." *Philadelphia Inquirer*, January 17.

Pollier, Fred S. 1989. "Resource Recovery in the 90s: A New Realism." *Resource Recovery*, April.

Popp, Paul O., Norman L. Hecht, and Rick E. Melberth. 1985. *Decision-Making in Local Government: The Resource Recovery Alternative*. Lancaster, Pa.: Technomic Publishing Co.

Pulver, Ellen. 1986. "State Needs Trash Plants, Experts Say." *Philadelphia Inquirer*, April 2.

Purifico, Regina Ann. 1985a. "Residents Plan Protest of Trash Zone in Plymouth." *Norristown Times Herald*, April 17.

———. 1985b. "Bartle: Sought Cooperation for Trash Plan." *Norristown Times Herald*, April 25.

———. 1985c. "Legislators Irk TRASH Leader." *Norristown Times Herald*, May 31.

Purser, Mark. 1990a. "Lisbon Tells County: Don't Import Trash." *Ogdensburg Journal*, June 14.

———. 1990b. "Morse Raps Incinerator, Brydon: It's 'Criminal' to Dump Project." *Ogdensburg Journal*, June 15.

Quinn, Rose. 1987. "Leake Says City Getting Bums' Rush." *Delaware County Sunday Times*, February 27, p. 18.

Raymo, Denise. 1989. "SWDA to Doctors: Tell Us Why You Oppose Project." *Ogdensburg Journal*, March 20.

Reagen, Jim. 1989. "Rally Against Incinerator." *Plaindealer*, May 10.

———. 1990. "DEC Incinerator OK May Kill Project." *Advance*, June 10.

Ross, Elizabeth. 1992. "Community Recycling Reduces Trash Trail." *Christian Science Monitor*, June 18, p. 11.

Rossie, David. 1989. "Legislators Try to Avoid Burning Issue." *Broome County Press & Sun-Bulletin*, November 12.

———. 1991. "Recycling Is Incinerator's Enemy." *Broome County Press & Sun-Bulletin*, November 4.

Roy, Yancey. 1991. "Protest: Hundreds Rally to Seek Funds for Recycling Programs." *Albany Times Union*, March 5.

Sbarra, Don. 1988a. "Kirkwood Offered $6 Million Incinerator Consolation." *Broome County Press & Sun-Bulletin*, July 6.

———. 1988b. "Broome Turns to State Court to Control Land." *Broome County Press & Sun-Bulletin*, August 4.

———. 1988c. "Democrats Try to Shield Incinerator Project from Grippen." *Broome County Press & Sun-Bulletin*, November 18.

———. 1989. "Refusal Leads to Oath Argument." *Broome County Press & Sun-Bulletin*, July 3.

———. 1990. "DEC Judge Douses Burner Review." *Broome County Press & Sun-Bulletin*, November 3.

———. 1991a. "Broome Lawmakers Override Incinerator Aid Veto." *Broome County Press & Sun-Bulletin*, February 22.

———. 1991b. "A Burning Issue: Hearings Begin Today on Broome County's Incinerator Plans." *Broome County Press & Sun-Bulletin*, June 11.

———. 1991c. "Burner Debate Outlasts Hearing." *Broome County Press & Sun-Bulletin*, June 28.

———. 1992a. "Trash Imports OK'd: Legislative Incinerator Vote Veto-Proof." *Broome County Press & Sun-Bulletin*, February 7.

———. 1992b. "Burner Builder Stands By: Foster Wheeler Tells Broome It Should Pursue Smaller Incinerator." *Broome County Press & Sun-Bulletin*, February 25.

———. 1993a. "Incinerator Builder Sues: Grippen Confident County Will Beat $27 Million Demand." *Broome County Press & Sun-Bulletin*, February 26.

———. 1993b. "Broome Takes Out Trash: 1992 Report Shows County Recycled, Reduced Nearly Half of Expected Waste." *Broome County Press & Sun-Bulletin*, May 2.

Schnaiberg, Allan, and Kenneth Gould. 1994. *Environment and Society: The Enduring Conflict*. New York: St. Martin's Press.

Selke, Susan E. 1990. *Packaging and the Environment: Alternatives, Trends, and Solutions*. Lancaster, Pa.: Technomic Publishing Co.

Shirm, G., J. Nash, and J. Balter. 1986. "City's Trash Should Be Recycled." *Philadelphia Inquirer*, May 19.

Smith, Michael. 1987. "Incinerator Construction Start Expected by June." *Sunday News*, March 29.

Snow, David A., and Robert Benford. 1992. "Master Frames and Cycles of Protest." In Aldon Morris and Carol McClurg Mueller, eds., *Frontiers in Social Movement Theory*, pp. 133–55. New Haven, Conn.: Yale University Press.

Snow, David A., Daniel Cress, Liam Downey, and Andrew Jones. 1996. "Disrupting the 'Quotidian': Reconceptualizing the Relationship Between Breakdown and Collective Action." Unpublished paper.

Snow, David A., E. Burke Rochford, Stephen Worden, and Robert Benford. 1986. "Frame Alignment and Mobilization." *American Sociological Review* 51:464–81.

Stranahan, Susan Q. 1995. "Now, It Seems, Trash Is Worth Fighting Over." *Philadelphia Inquirer*, December 10.

Sutton, W. W. 1987. "Chester Mayor Tells of Goode's Visit." *Philadelphia Inquirer*, February 7.

Sutton, W. W., Mark Jaffe, and Russell Cooke. 1985. "Council Rejects Trash-to-Steam." *Philadelphia Inquirer*, January 11.

Szasz, Andrew. 1994. *EcoPopulism: Toxic Waste and the Movement for Environmental Justice*. Minneapolis: University of Minnesota Press.

Tarrow, Sidney. 1994. *Power in Movement*. New York: Cambridge University Press.

Taylor, Adam. 1988a. "McKellar: Let Westy Build." *Delaware County Daily Times*, April 25.

———. 1988b. "Chester Official's Letter Steams Westinghouse." *Delaware County Daily Times*, April 27.

———. 1988c. "Westy, Delco Sweeten City Trash Deal." *Delaware County Daily Times*, June 17.

Tilly, Charles. 1978. *From Mobilization to Revolution*. Reading, Mass.: Addison-Wesley.

Tooley, Shawn G. 1990. " 'Foreign' Trash Hot Issue: Legislature, SWDA Disagree on Importation Powers." *Watertown Daily Times*, May 21.

U.S. Congress, Office of Technology Assessment (OTA). 1989. *Facing America's Trash: What Next for Municipal Solid Waste?* OTA-0-424. Washington, D.C.: U.S. Government Printing Office.

Useem, Bert. 1980. "Solidarity Model, Breakdown Model, and the Boston Anti-Busing Movement." *American Sociological Review* 45:357–69.

Vis, David. 1988. "DEP: Mass Burn Can't Be Delayed." *Atlantic City Press*, February 6.

Walsh, Edward. 1987. "Challenging Official Risk Assessments via Protest Mobilization: the TMI Case." In B. B. Johnson and V. T. Covello, eds., *The Social and Cultural Construction of Risk*, pp. 85–102. Dordrecht: Reidel.

———. 1988a. "New Dimensions of Social Movements: The High-Level Waste-Siting Controversy." *Sociological Forum* 3(4): 586–605.

———. 1988b. *Democracy in the Shadows: Citizen Mobilization in the Wake of the Accident at Three Mile Island*. Westport, Conn.: Greenwood Press.

Walsh, Edward, and Rex Warland. 1983. "Social Movement Involvement in the Wake of a Nuclear Accident." *American Sociological Review* 48:764–80.

Walsh, Edward, Rex Warland, and D. Clayton Smith. 1993. "Backyards, NIMBYs, and Incinerator Sitings: Implications for Social Movement Theory." *Social Problems* 40(1): 25–38.

Watson, R. Pete. 1988. *Resource Recovery Newsletter*, January.

Weiss, Todd. 1988. "Reconsider Trash-to-Steam Plant, Consultant Says." *Norristown Times Herald*, November 22.

Whitaker, Jennifer S. 1994. *Salvaging the Land of Plenty: Garbage and the American Dream*. New York: William Morrow & Co.

Wiegand, Ginny. 1986. "They Take Democracy Personally." *Philadelphia Inquirer*, January 9.

York County Planning Commission. 1985. "York County Municipalities Solid Waste Management Plan Update: Final Report (December 1985)." Unpublished.

Zald, Mayer. 1992. "Looking Backward to Look Forward: Reflections on the Past and Future of the Resource Mobilization Program." In Aldon D. Morris and Carol McClurg Mueller, eds., *Frontiers in Social Movement Theory*, pp. 326–48. New Haven, Conn.: Yale University Press.

Zald, Mayer, and Bert Useem. 1987. "Movement and Countermovement Interaction." In Mayer Zald and John McCarthy, eds., *Social Movements in an Organizational Society*, pp. 247–72. New Brunswick, N.J.: Transaction Books.

Zelnik, Joe. 1987a. "Burn Foes Win Delay in MUA Incinerator." *Cape May County Herald-Lantern-Dispatch*, March 18.

———. 1987b. "Thornton, Frederick Call MUA, Foes 'Extremists.' " *Cape May County Herald-Lantern-Dispatch*, May 20.

Zissu, Erik M. 1989. "Medical Society Endorses Incinerator Project Delay." *Watertown Daily Times*, March 18.

Index